Learning SciPy for Numerical and Scientific Computing
Second Edition

Quick solutions to complex numerical problems in physics, applied mathematics, and science with SciPy

Sergio J. Rojas G.

Erik A Christensen

Francisco J. Blanco-Silva

BIRMINGHAM - MUMBAI

Learning SciPy for Numerical and Scientific Computing
Second Edition

First published: February 2013

Second edition: February 2015

Production reference: 1200215

Published by Packt Publishing Ltd.
Livery Place
35 Livery Street
Birmingham B3 2PB, UK.

ISBN 978-1-78398-770-2

www.packtpub.com

Credits

Authors
Sergio J. Rojas G.
Erik A Christensen
Francisco J. Blanco-Silva

Reviewers
Dr. Robert Clewley
Nicolas Fauchereau
Valentin Haenel
Andy Ray Terrel

Commissioning Editor
Kartikey Pandey

Acquisition Editors
Kartikey Pandey
Meeta Rajani

Content Development Editor
Shweta Pant

Technical Editor
Rahul C. Shah

Copy Editors
Roshni Banerjee
Puja Lalwani
Merilyn Pereira

Project Coordinator
Shipra Chawhan

Proofreaders
Paul Hindle
Clyde Jenkins

Indexers
Monica Ajmera Mehta
Priya Sane

Graphics
Sheetal Aute
Valentina D'silva
Abhinash Sahu

Production Coordinator
Nitesh Thakur

Cover Work
Nitesh Thakur

About the Authors

Sergio J. Rojas G. is currently a full professor of physics at Universidad Simón Bolívar, Venezuela. Regarding his formal studies, in 1991, he earned a BS in physics with his thesis on numerical relativity from the Universidad de Oriente, Estado Sucre, Venezuela, and then, in 1998, he earned a PhD in physics from the Department of Physics at City College of the City University of New York, where he worked on the applications of fluid dynamics in the flow of fluids in porous media, gaining and developing since then a vast experience in programming as an aid to scientific research via Fortran77/90 and C/C++. In 2001, he also earned a master's degree in computational finance from the Oregon Graduate Institute of Science and Technology.

Sergio's teaching activities involve lecturing undergraduate and graduate physics courses at his home university, Universidad Simón Bolívar, Venezuela, including a course on Monte Carlo methods and another on computational finance. His research interests include physics education research, fluid flow in porous media, and the application of the theory of complex systems and statistical mechanics in financial engineering. More recently, Sergio has been involved in machine learning and its applications in science and engineering via the Python programming language.

I am deeply grateful to my mother, Eufemia del Valle Rojas González, a beloved woman whose given steps were always in favor of showing and upraising the best of a human being.

Erik A Christensen is a quant analyst/developer in finance and creative industries. He has a PhD from the Technical University of Denmark, with postdoctoral studies at the Levich Institute at the City College of the City University of New York and the Courant Institute of Mathematical Sciences at New York University. His interests in technology span from Python to F# and Cassandra/Spark. He is active in the meet-up communities in London!

I would like to thank my family and friends for their support during this work!

Francisco J. Blanco-Silva is the owner of a scientific consulting company—Tizona Scientific Solutions—and adjunct faculty in the Department of Mathematics of the University of South Carolina. He obtained his formal training as an applied mathematician at Purdue University. He enjoys problem solving, learning, and teaching. Being an avid programmer and blogger, when it comes to writing, he relishes finding that common denominator among his passions and skills and making it available to everyone. He coauthored *Modeling Nanoscale Imaging in Electron Microscopy*, *Springer* along with Peter Binev, Wolfgang Dahmen, and Thomas Vogt.

About the Reviewers

Dr. Robert Clewley is a polymath scientist and educator. He has been a faculty member at Georgia State University, Atlanta, GA. He specializes in computational and mathematical modeling methods for complex adaptive systems and has published a diverse range of academic journals involving applications in epilepsy, cancer, cardiology, and biomechanics. His research has been supported by federal grants from NSF and the Army Research Laboratory. From the high school level to graduate degree level, he has developed and taught a variety of courses spanning mathematics, computer science, physics, biological sciences, and philosophy of science. Dr. Clewley also develops the open source PyDSTool modeling software that is used internationally in many scientific and engineering fields.

Nicolas Fauchereau is a climate scientist at the National Institute for Water and Atmospheric Research (NIWA Ltd.) based in Auckland, New Zealand.

After obtaining his PhD in France in 2004, he spent 7 years in South Africa working at the University of Cape Town and then at the Council for Scientific and Industrial Research, before joining NIWA in 2012.

He uses statistics, data mining, and machine learning to try and make sense of climate and environmental data and to develop solutions to help people anticipate and adapt to climate variability and change.

He's been using the Python scientific stack for about 10 years and is a passionate advocate for the use of Python in environmental and earth sciences.

A water sports enthusiast, he likes to spend his free time either surfing, kite surfing, or sailing with his wife and two kids.

Valentin Haenel is a software engineer interested in the architectures of high-performance number crunching with Python. Specifically, he is interested in low-level aspects such as interfacing Python with C code, strategies for efficient memory allocation, avoiding redundant memory copies, and exploiting the memory hierarchy for accelerated computation. He spends some of his spare time working on Blosc (`http://blosc.org`), an extremely fast and multi threaded meta-codec. Occasionally, he flirts with machine learning.

In the past, he had worked on psychophysics data analysis, large-scale brain simulations, analytical engines for business intelligence, and large-scale data-center monitoring. He wrote a book about using the Git version control system and has contributed to a diverse selection of over 50 open source projects. He currently resides in Berlin and works as a freelance software engineer, consultant, and trainer.

www.PacktPub.com

Support files, eBooks, discount offers, and more

For support files and downloads related to your book, please visit www.PacktPub.com.

Did you know that Packt offers eBook versions of every book published, with PDF and ePub files available? You can upgrade to the eBook version at www.PacktPub.com and as a print book customer, you are entitled to a discount on the eBook copy. Get in touch with us at service@packtpub.com for more details.

At www.PacktPub.com, you can also read a collection of free technical articles, sign up for a range of free newsletters and receive exclusive discounts and offers on Packt books and eBooks.

https://www2.packtpub.com/books/subscription/packtlib

Do you need instant solutions to your IT questions? PacktLib is Packt's online digital book library. Here, you can search, access, and read Packt's entire library of books.

Why subscribe?

- Fully searchable across every book published by Packt
- Copy and paste, print, and bookmark content
- On demand and accessible via a web browser

Free access for Packt account holders

If you have an account with Packt at www.PacktPub.com, you can use this to access PacktLib today and view 9 entirely free books. Simply use your login credentials for immediate access.

Table of Contents

Preface

While maintaining the main structure of the first edition, this revised edition of *Learning SciPy for Numerical and Scientific Computing* includes a set of companion IPython Notebooks. This will help students, researchers, and practitioners modify and incorporate in their own work, the set of tested code snippets that are presented in the book, as the pedagogical strategy. This will also show and illustrate the computing power that SciPy brings to the fingertips of anyone interested in performing numerical computation via the unique flexibility offered by the Python computer language.

We should mention, however, that the IPython Notebooks will make sense to anyone starting in the field only if they are read alongside the corresponding section in the book, helping you to develop skills in the use of SciPy to solve large scale numerical problems while gaining understanding of the conditions and limitations associated with the modules contained in SciPy. Certainly, the already knowledgeable reader will find pleasure as they encounter material they already know, but will be challenged to devise better ways to accomplish with the same level of clarity presented in the book with the many computational tasks used to illustrate the functionality of SciPy.

SciPy has been an integral part of the computational environment of choice for many scientists for years. One of our challenges today is to bring together professionals with different backgrounds, technologies, and expertise in software (from the pure mathematician, to the hardcore engineer) to contribute independent of their working environments.

SciPy in Python is a perfect platform to coordinate projects in a smooth, reliable, and coherent environment. It allows performing most tasks with ease; reason being that many dedicated software tools easily integrate with the core features of SciPy, therefore, interfacing with non-Python-based software packages and tools is becoming increasingly simple.

In summary, this book presents the most robust programming environment to date. We will show you how to use this system from basic manipulation of data, to a very detailed exposition through examples in different branches of science and engineering.

What this book covers

Chapter 1, Introduction to SciPy, shows the benefits of using the combination of Python, NumPy, SciPy, and matplotlib as a programming environment for scientific purposes. You will learn how to install, test, and explore the environments, use them for quick computations, and figure out a few good ways to search for help. A brief introduction on how to open the companion IPython Notebooks that comes with this book is also presented.

Chapter 2, Working with the NumPy Array As a First Step to SciPy, explores in depth the creation and basic manipulation of the object array used by SciPy, as an overview of the NumPy libraries.

Chapter 3, SciPy for Linear Algebra, covers applications of SciPy to applications with large matrices, including solving systems or computation of eigenvalues and eigenvectors.

Chapter 4, SciPy for Numerical Analysis, is without a doubt one of the most interesting chapters in this book. It covers with great detail the definition and manipulation of functions (one or several variables), the extraction of their roots, extreme values (optimization), computation of derivatives, integration, interpolation, regression, and applications to the solution of ordinary differential equations.

Chapter 5, SciPy for Signal Processing, explores construction, acquisition, quality improvement, compression, and feature extraction of signals (in any dimension). It is covered with beautiful and interesting examples from the field of image processing.

Chapter 6, SciPy for Data Mining, covers applications of SciPy for collection, organization, analysis, and interpretation of data, with examples taken from statistics and clustering.

Chapter 7, SciPy for Computational Geometry, explores the construction of triangulation of points, convex hulls, Voronoi diagrams, and applications, including the solving of the two dimensional Laplace Equation via the Finite Element Method in a rectangular grid. At this point in the book, it will be possible to combine techniques from all the previous chapters to show state-of-the-art research performed with ease with SciPy, and we will explore a few good examples from Material Science and Experimental Physics.

Chapter 8, Interaction with Other Languages, introduces one of the main strengths of SciPy — the ability to interact with other languages such as C/C++, Fortran, R, and MATLAB/Octave.

What you need for this book

To work with the examples and try out the code in this book, all you need is a recent build of Python (2.7 or higher) with the libraries NumPy, SciPy, and matplotlib. Recipes to install all these are provided throughout the book.

Who this book is for

This book is for scientists, engineers, programmers, or analysts with knowledge of Python. For some of the sections, a decent command over linear algebra, calculus, and some statistics is needed to understand some of the concepts, but otherwise this book is mostly self-contained.

Conventions

In this book, you will find a number of styles of text that distinguish between different kinds of information. Here are some examples of these styles, and an explanation of their meaning.

Code words in text, database table names, folder names, filenames, file extensions, pathnames, dummy URLs, user input, and Twitter handles are shown as follows: "We can include other contexts through the use of the include directive."

A block of code is set as follows:

```
solve(A, b, sym_pos=False, lower=False, overwrite_a=False,
overwrite_b=False, debug=False)
spsolve(A, b[, permc_spec, use_umfpack])
```

The reader with the required background should recognize the Python prompt >>> followed by a space and then the code field. Any command-line input or output is written as follows:

```
>>> from scipy import stats
>>> result=scipy.stats.bayes_mvs(scores)
```

New terms and **important words** are shown in bold. Words that you see on the screen, in menus or dialog boxes for example, appear in the text like this: "clicking the **Next** button moves you to the next screen".

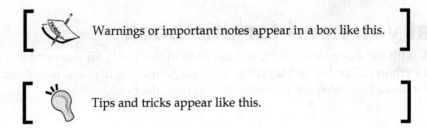

> Warnings or important notes appear in a box like this.

> Tips and tricks appear like this.

Reader feedback

Feedback from our readers is always welcome. Let us know what you think about this book—what you liked or may have disliked. Reader feedback is important for us to develop titles that you really get the most out of.

To send us general feedback, simply send an e-mail to feedback@packtpub.com, and mention the book title via the subject of your message.

If there is a topic that you have expertise in and you are interested in either writing or contributing to a book, see our author guide on www.packtpub.com/authors.

Customer support

Now that you are the proud owner of a Packt book, we have a number of things to help you to get the most from your purchase.

Downloading the example code

You can download the example code files for all Packt books you have purchased from your account at http://www.packtpub.com. If you purchased this book elsewhere, you can visit http://www.packtpub.com/support and register to have the files e-mailed directly to you.

Downloading the color images of this book

We also provide you a PDF file that has color images of the screenshots/diagrams used in this book. The color images will help you better understand the changes in the output. You can download this file from `https://www.packtpub.com/sites/default/files/downloads/7702OS_ColoredImages.pdf`.

Errata

Although we have taken every care to ensure the accuracy of our content, mistakes do happen. If you find a mistake in one of our books—maybe a mistake in the text or the code—we would be grateful if you would report this to us. By doing so, you can save other readers from frustration and help us improve subsequent versions of this book. If you find any errata, please report them by visiting `http://www.packtpub.com/submit-errata`, selecting your book, clicking on the **errata submission form** link, and entering the details of your errata. Once your errata are verified, your submission will be accepted and the errata will be uploaded on our website, or added to any list of existing errata, under the Errata section of that title. Any existing errata can be viewed by selecting your title from `http://www.packtpub.com/support`.

Piracy

Piracy of copyright material on the Internet is an ongoing problem across all media. At Packt, we take the protection of our copyright and licenses very seriously. If you come across any illegal copies of our works, in any form, on the Internet, please provide us with the location address or website name immediately so that we can pursue a remedy.

Please contact us at `copyright@packtpub.com` with a link to the suspected pirated material.

We appreciate your help in protecting our authors, and our ability to bring you valuable content.

Questions

You can contact us at `questions@packtpub.com` if you are having a problem with any aspect of the book, and we will do our best to address it.

1
Introduction to SciPy

There is no doubt that the labor of scientists in the twenty-first century is more comprehensive and interdisciplinary than in previous generations. Members of scientific communities connect in larger teams and work together on mission-oriented goals and across their fields. This paradigm on research is also reflected in the computational resources employed by researchers. No longer are researchers restricted to one type of commercial software, operating system, or vendor, but inspired by open source contributions made available and tested by research institutions and open source communities; research work often spans over various platforms and technologies.

This book presents the highly-recognized open source programming environment till date — a system based on two libraries of the computer language Python: **NumPy** and **SciPy**. In the following sections, we will guide you through examples from science and engineering on the usage of this system.

What is SciPy?

The ideal programming environment for computational mathematics enjoys the following characteristics:

- It must be based on a computer language that allows the user to work quickly and integrate systems effectively. Ideally, the computer language should be portable to all platforms: Windows, Mac OS X, Linux, Unix, Android, and so on. This is key to fostering cooperation among scientists with different resources and accessibilities. It must contain a powerful set of libraries that allow the acquisition, storing, and handling of large datasets in a simple and effective manner. This is central — allowing simulation and the employment of numerical computations at a large scale.

- Smooth integration with other computer languages, as well as third-party software.

- Besides running the compiled code, the programming environment should allow the possibility of interactive sessions as well as scripting capabilities for quick experimentation.

- Different coding paradigms should be supported — imperative, object-oriented, and/or functional coding styles.

- It should be an open source software, that allows user access to the raw data code, and allows the user to modify basic algorithms if so desired. With commercial software, the inclusion of the improved algorithms is applied at the discretion of the seller, and it usually comes at a cost of the end user. In the open source universe, the community usually performs these improvements and releases new versions as they are published — at no cost.

- The set of applications should not be restricted to mere numerical computations; it should be powerful enough to allow symbolic computations as well.

Among the best-known environments for numerical computations used by the scientific community is **MATLAB**, which is commercial, expensive, and which does not allow any tampering with the code. **Maple** and **Mathematica** are more geared towards symbolic computation, although they can match many of the numerical computations from MATLAB. These are, however, also commercial, expensive, and closed to modifications. A decent alternative to MATLAB and based on a similar mathematical engine is the **GNU Octave system**. Most of the MATLAB code is easily portable to Octave, which is open source. Unfortunately, the accompanying programming environment is not very user friendly, it is also very much restricted to numerical computations. One environment that combines the best of all worlds is Python with the open source libraries NumPy and SciPy for numerical operations. The first property that attracts users to Python is, without a doubt, its code readability. The syntax is extremely clear and expressive. It has the advantage of supporting code written in different paradigms: object oriented, functional, or old school imperative. It allows packing of Python codes and to run them as standalone executable programs through the `py2exe`, `pyinstaller`, and `cx_Freeze` libraries, but it can also be used interactively or as a scripting language. This is a great advantage when developing tools for symbolic computation. Python has therefore been a firm competitor to Maple and Mathematica: the open source mathematics software **Sage (System for Algebra and Geometry Experimentation)**.

NumPy is an open source extension to Python that adds support for multidimensional arrays of large sizes. This support allows the desired acquisition, storage, and complex manipulation of data mentioned previously. NumPy alone is a great tool to solve many numerical computations.

On top of NumPy, we have yet another open source library, SciPy. This library contains algorithms and mathematical tools to manipulate NumPy objects with very definite scientific and engineering objectives.

The combination of Python, NumPy, and SciPy (which henceforth are coined as "SciPy" for brevity) has been the environment of choice of many applied mathematicians for years; we work on a daily basis with both pure mathematicians and with hardcore engineers. One of the challenges of this trade is to bring about the scientific production of professionals with different visions, techniques, tools, and software to a single workstation. SciPy is the perfect solution to coordinate computations in a smooth, reliable, and coherent manner.

Constantly, we are required to produce scripts with, for example, combinations of experiments written and performed in SciPy itself, C/C++, Fortran, and/or MATLAB. Often, we receive large amounts of data from some signal acquisition devices. From all this heterogeneous material, we employ Python to retrieve and manipulate the data, and once finished with the analysis, to produce high-quality documentation with professional-looking diagrams and visualization aids. SciPy allows performing all these tasks with ease.

This is partly because many dedicated software tools easily extend the core features of SciPy. For example, although graphing and plotting are usually taken care of with the Python libraries of **matplotlib**, there are also other packages available, such as **Biggles** (http://biggles.sourceforge.net/), **Chaco** (https://pypi.python.org/pypi/chaco), **HippoDraw** (https://github.com/plasmodic/hippodraw), **MayaVi** for **3D** rendering (http://mayavi.sourceforge.net/), the **Python Imaging Library** or **PIL** (http://pythonware.com/products/pil/), and the online analytics and data visualization tool **Plotly** (https://plot.ly/).

Interfacing with non-Python packages is also possible. For example, the interaction of SciPy with the R statistical package can be done with **RPy** (http://rpy.sourceforge.net/rpy2.html). This allows for much more robust data analysis.

Installing SciPy

At the time of this book, the stable production releases of Python were 2.7.9 and 3.4.2. Still, Python 2.7 is more convenient if the user needs to communicate with third-party applications. No new releases are planned for Python 2; Python 3 is considered the present and the future of Python. For the purposes of SciPy applications, we do recommend you hold on to the 2.7 version, as there are still some packages using SciPy that have not been ported to Python 3 yet. Nevertheless, the companion software of this book was tested to work on both Python 2.7 and Python 3.4.

The Python software package can be downloaded from the official site (`https://www.python.org/downloads/`) and can be installed on all major systems such as Windows, Mac OS X, Linux, and Unix. It has also been ported to other platforms, including Palm OS, iOS, PlayStation, PSP, Psion, and so on.

The following screenshot shows two popular options for coding in Python on an iPad—**PythonMath** and **Sage Math**. While the first application allows only the use of simple math libraries, the second permits the user to load and use both NumPy and SciPy remotely.

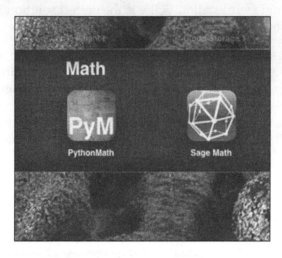

PythonMath and **Sage Math** bring Python coding to iOS devices. **Sage Math** allows importing NumPy and SciPy.

We shall not go into detail about the installation of Python on your system, since we already assume familiarity with this language. In case of doubt, we advise browsing the excellent book *Expert Python Programming*, *Tarek Ziadé, Packt Publishing*, where detailed explanations are given for installing many of the different implementations on different systems. It is usually a good idea to follow the directions given on the official Python website. We will also assume familiarity with carrying out interactive sessions in Python, as well as writing standalone scripts.

The latest libraries for both NumPy and SciPy can be downloaded from the official SciPy site (`http://scipy.org/`). They both require a Python Version 2.4 or newer, so we should be in good shape at this point. We may choose to download the package from SourceForge (`http://sourceforge.net/projects/scipy/`), **Gohlke** (`http://www.lfd.uci.edu/~gohlke/pythonlibs/`) or **Git** repositories (for instance, the **superpack** from `http://stronginference.com/ScipySuperpack/`).

It is also possible in some systems to use prepackaged executable bundles that simplify the process, such as the **Anaconda** (https://store.continuum.io/cshop/anaconda/) or the **Enthought** (https://www.enthought.com/products/epd/) Python distributions. Here, we will show you how to download and install Scipy on various platforms in the most common cases.

Installing SciPy on Mac OS X

While installing SciPy on Mac OS X, you must consider some criteria before you install it on your system. This helps in smooth installation of SciPy. The following are the things to be taken care of:

- For instance, in Mac OS X, if `MacPorts` is installed, the process could not be easier. Open a terminal as superuser, and at the prompt (%), issue the following command:

```
% port search scipy
```

- This presents a list of all ports that either install SciPy or use SciPy as a requirement. For Python 2.7 we need to install `py27-scipy` issuing the following command:

```
% port install py27-scipy
```

A few minutes later, the libraries are properly installed and ready to use. Note how `macports` also installs all needed requirements for us (including the NumPy libraries) without any extra effort on our part.

Installing SciPy on Unix/Linux

Under any other Unix/Linux system, if either no ports are available or if the user prefers to install from the packages downloaded from either SourceForge or Git, it is enough to perform the following steps:

1. Unzip the NumPy and SciPy packages following the recommendation of the official pages. This creates two folders, one for each library.

   Within a terminal session, change directories to the folder where the NumPy libraries are stored, which contains the `setup.py` file. Find out which Fortran compiler you are using (one of `gnu`, `gnu95`, or `fcompiler`), and at prompt, issue the following command:

   ```
   % python setup.py build –fcompiler=<compiler>
   ```

2. Once built, and on the same folder, issue the installation command. This should be all:

```
% python setup.py install
```

Installing SciPy on Windows

You can install Scipy on Windows in many ways. The following are some recommended ways that you might want to have a look on:

- Under Microsoft Windows, we recommend you install from the binary installers provided by the Anaconda or Enthought Python Distributions. Please, however, be aware of the memory requirements. Alternatively, you can download and install the SciPy stack or the libraries, individually.

- The procedure for the installation of the SciPy libraries is exactly the same, that is, downloading and building before installing under Unix/Linux or downloading and running under Microsoft Windows. Note that different implementations of Python might have different requirements before installing NumPy and SciPy.

Testing the SciPy installation

As you might know, computer systems are not infallible. Accordingly, before starting computing via SciPy, one needs to be sure it is working correctly. To that end, SciPy developers have included a test suit any user of SciPy can execute to be sure the SciPy being used is working fine. That way, much debugging time can be saved whenever an error occurs while using any function provided by SciPy.

To run the test suite, at the Python prompt, one can run the following commands:

```
>>> import scipy
>>> scipy.test()
```

> **Downloading the example code**
>
> You can download the example code files for all Packt books you have purchased from your account at http://www.packtpub.com. If you purchased this book elsewhere, you can visit http://www.packtpub.com/support and register to have the files e-mailed directly to you.

The reader should be aware that the execution of this test will take some time to finish. It should end with something like this:

```
File Edit View Search Terminal Help
>>> import scipy
>>> scipy.test()
Running unit tests for scipy
NumPy version 1.8.0
NumPy is installed in /opt/python2.76/site-packages/numpy
SciPy version 0.13.3
SciPy is installed in /opt/python2.76/site-packages/scipy
Python version 2.7.6 [GCC 4.1.2 20080704 (Red Hat 4.1.2-54)]
nose version 1.3.0
...
...
...
----------------------------------------------------------------------
Ran 8936 tests in 194.730s

OK (KNOWNFAIL=115, SKIP=204)
<nose.result.TextTestResult run=8936 errors=0 failures=0>
>>>
```

This means that at the basic level, your SciPy installation is fine. Eventually, the test could end in the form:

```
File Edit View Search Terminal Help
>>> import scipy
>>> scipy.test()
Running unit tests for scipy
NumPy version 1.9.0
NumPy is installed in /opt/python3.41/site-packages/numpy
SciPy version 0.14.0
SciPy is installed in /opt/python3.41/site-packages/scipy
Python version 3.4.1 [GCC 4.4.7 20120313 (Red Hat 4.4.7-1)]
nose version 1.3.4
...
...
...
----------------------------------------------------------------------
Ran 16413 tests in 363.062s

FAILED (KNOWNFAIL=277, SKIP=904, errors=326, failures=45)
<nose.result.TextTestResult run=16413 errors=326 failures=45>
>>>
```

In this case, one needs to revise carefully the errors and the failed tests. A place to get help is the SciPy mailing list (http://mail.scipy.org/pipermail/scipy-user/) to which one could subscribe. We have included a Python script that the reader could use to run these tests that can be found at the companion software for this chapter that comes with the book.

SciPy organization

SciPy is organized as a family of modules. We like to think of each module as a different field of mathematics. And as such, each has its own particular techniques and tools. You can find a list of some of the different modules included in SciPy at `http://docs.scipy.org/doc/scipy-0.14.0/reference/py-modindex.html`.

Let's use some of its functions to solve a simple problem.

The following table shows the IQ test scores of 31 individuals:

114	100	104	89	102	91	114	114
103	105	108	130	120	132	111	128
118	119	86	72	111	103	74	112
107	103	98	96	112	112	93	

A stem plot of the distribution of these 31 scores (refers to the IPython Notebook for this chapter) shows that there are no major departures from normality, and thus we assume the distribution of the scores to be close to normal. Now, estimate the mean IQ score for this population, using a 99 percent confidence interval.

We start by loading the data into memory, as follows:

```
>>> import numpy
>>> scores = numpy.array([114, 100, 104, 89, 102, 91, 114, 114, 103, 105,
108, 130, 120, 132, 111, 128, 118, 119, 86, 72, 111, 103, 74, 112, 107,
103, 98, 96, 112, 112, 93])
```

At this point, if we type `dir(scores)`, hit the *return* key followed by a dot (`.`), and press the *tab* key ;the system lists all possible methods inherited by the data from the NumPy library, as it is customary in Python. Technically, we could go ahead and compute the required `mean`, `xmean`, and corresponding confidence interval according to the formula, *xmean ± zcrit * sigma / sqrt(n)*, where `sigma` and `n` are respectively the standard deviation and size of the data, and *zcrit* is the critical value corresponding to the confidence (`http://en.wikipedia.org/wiki/Confidence_interval`). In this case, we could look up a table on any statistics book to obtain a crude approximation to its value, *zcrit* = 2.576. The remaining values may be computed in our session and properly combined, as follows:

```
>>> import scipy
>>> xmean = scipy.mean(scores)
>>> sigma = scipy.std(scores)
```

```
>>> n = scipy.size(scores)
>>> xmean, xmean - 2.576*sigma /scipy.sqrt(n), \
    xmean + 2.576*sigma / scipy.sqrt(n)
```

The output is shown as follows:

```
(105.83870967741936, 99.343223715529746, 112.33419563930897)
```

We have thus computed the estimated mean IQ score (with value `105.83870967741936`) and the interval of confidence (from about `99.34` to approximately `112.33`). We have done so using purely SciPy-based operations while following a known formula. But instead of making all these computations by hand and looking for critical values on tables, we could just ask SciPy.

Note how the `scipy.stats` module needs to be loaded before we use any of its functions:

```
>>> from scipy import stats
>>> result=scipy.stats.bayes_mvs(scores)
```

The variable `result` contains the solution to our problem with some additional information. Note that result is a tuple with three elements as the `help` documentation suggests:

```
>>> help(scipy.stats.bayes_mvs)
```

The output of this command will depend on the installed version of SciPy. It might look like this (run the companion IPython Notebook for this chapter to see how the actual output from your system is, or run the command in a Python console):

```
Help on function bayes_mvs in module scipy.stats.morestats:

bayes_mvs(data, alpha=0.90000000000000002)
    Return Bayesian confidence intervals for the mean, var, and std.

    Assumes 1-d data all has same mean and variance and uses Jeffrey's prior
    for variance and std.

    alpha gives the probability that the returned confidence interval contains
    the true parameter.

    Uses mean of conditional pdf as center estimate
    (but centers confidence interval on the median)

    Returns (center, (a, b)) for each of mean, variance and standard deviation.
    Requires 2 or more data-points.
(END)
```

Our solution is the first element of the tuple `result`; to see its contents, type:

```
>>> result[0]
```

The output is shown as follows:

```
(105.83870967741936, (101.48825534263035, 110.18916401220837))
```

Note how this output gives us the same average as before, but a slightly different confidence interval, due to more accurate computations through SciPy (the output might be different depending on the SciPy version available on your computer).

How to find documentation

There is a wealth of information online, either from the official pages of SciPy (although its reference guides are somehow incomplete, as a work in progress), or from many other contributors that present tutorials on forums, YouTube, or personal sites. Several developers also publish examples of their work with great detail online.

As we previously saw, it is also possible to obtain help from our interactive Python sessions. The libraries NumPy and SciPy make use of **docstrings** heavily, which makes it simple to request for help for usage and recommendations with the usual Python help system. For example, if in doubt of the usage of the `bayes_mvs` routine, the user can issue the following command:

```
>>> import scipy.stats
>>> help(scipy.stats.bayes_mvs)
```

After executing this command, the system provides the necessary information. Equivalently, both NumPy and SciPy come bundled with their own help system, `info`. For instance, look at the following command:

```
>>> import numpy
>>> numpy.info('random')
```

This will offer a summary of all information parsed from the contents of all docstrings from the NumPy library associated with the given keyword (note it must be quoted). The user may navigate the output scrolling up and down, without the possibility of further interaction.

This is convenient provided we already do know the function we want to use if we are unsure of its usage. But, what should we do if we don't know about the existence of this procedure, and suspect that it may exist? The usual Python way is to invoke the `dir()` command on a module, which lists all possible attributes.

Interactive Python sessions make it easier to search for such information with the possibility of navigating and performing further searches inside the output of help sessions. For instance, type in the following command at prompt:

```
>>> import scipy.stats
>>> help(scipy.stats)
```

The output of this command will depend on the installed version of SciPy. It might look like this (run the companion IPython Notebook for this chapter to see the actual output from your system, or run the command in a Python console):

```
Help on package scipy.stats in scipy:

NAME
    scipy.stats

FILE
    /Applications/sage/local/lib/python2.6/site-packages/scipy/stats/__init__.py

DESCRIPTION
    Statistical Functions
    =====================

    This module contains a large number of probability distributions as
    well as a growing library of statistical functions.

    Each included distribution is an instance of the class rv_continous.
    For each given name the following methods are available.  See docstring for
    rv_continuous for more information

    :rvs:
        random variates with the distribution
    :pdf:
        probability density function
    :cdf:
        cumulative distribution function
    :sf:
        survival function (1.0 - cdf)
    :ppf:
        percent-point function (inverse of cdf)
    :isf:
        inverse survival function
    :stats:
        mean, variance, and optionally skew and kurtosis

    Calling the instance as a function returns a frozen pdf whose shape,
    location, and scale parameters are fixed.

    Distributions
    -------------

    The distributions available with the above methods are:
:
```

Note the colon (:) at the end of the screen—this is an old-school prompt. The system is in stand-by mode, expecting the user to issue a command (in the form of a single key). This also indicates that there are a few more pages of help following the given text. If we intend to read the rest of the help file, we may press spacebar to scroll to the next page.

In this way, we can visit the following manual pages on this topic. It is also possible to navigate the manual pages scrolling one line of text at a time using the up and down arrow keys. When we are ready to quit the help session, we simply press (the keyboard letter) *Q*.

It is also possible to search the help contents for a given string. In that case, at the prompt, we press the (/) slash key. The prompt changes from a colon into a slash, and we proceed to input the keyword we would like to search for.

For example, is there a SciPy function that computes the **Pearson kurtosis** of a given dataset? At the slash prompt, we type in `kurtosis` and press *enter*. The help system takes us to the first occurrence of that string. To access successive occurrences of the string kurtosis, we press the *N* key (for next) until we find what we require. At that stage, we proceed to quit this help session (by pressing *Q*) and request more information on the function itself:

```
>>> help(scipy.stats.kurtosis)
```

The output of this command will depend on the installed version of SciPy. It might look like this (run the companion IPython Notebook for this chapter to see how the actual output from your system is, or run the command in a Python console):

```
Help on function kurtosis in module scipy.stats.stats:

kurtosis(a, axis=0, fisher=True, bias=True)
    Computes the kurtosis (Fisher or Pearson) of a dataset.

    Kurtosis is the fourth central moment divided by the square of the
    variance. If Fisher's definition is used, then 3.0 is subtracted from
    the result to give 0.0 for a normal distribution.

    If bias is False then the kurtosis is calculated using k statistics to
    eliminate bias coming from biased moment estimators

    Use kurtosistest() to see if result is close enough to normal.

    Parameters
    ----------
    a : array
        data for which the kurtosis is calculated
    axis : int or None
        Axis along which the kurtosis is calculated
    fisher : bool
        If True, Fisher's definition is used (normal ==> 0.0). If False,
        Pearson's definition is used (normal ==> 3.0).
    bias : bool
        If False, then the calculations are corrected for statistical bias.

    Returns
    -------
    kurtosis : array
        The kurtosis of values along an axis. If all values are equal,
        return -3 for Fisher's definition and 0 for Pearson's definition.

    References
    ----------
    [CRCProbStat2000]_ Section  2.2.25

    .. [CRCProbStat2000] Zwillinger, D. and Kokoska, S. (2000). CRC Standard
        Probablity and Statistics Tables and Formulae. Chapman & Hall: New
        York. 2000.
(END)
```

Scientific visualization

At this point, we would like to introduce you to another resource that we will be using to generate graphs, namely the matplotlib libraries. It may be downloaded from its official web page, `http://matplotlib.org/`, and installed following the standard Python commands. There is a good online documentation in the official web page, and we encourage the reader to dig deeper than the few commands that we will use in this book. For instance, the excellent monograph *Matplotlib for Python Developers*, *Sandro Tosi, Packt Publishing*, provides all that we would need and more. Other plotting libraries are available (commercial or otherwise that aim to very different and specific applications. The degree of sophistication and ease of use of matplotlib makes it one of the best options to generate graphics in scientific computing.

Once installed, it may be imported using `import matplotlib`. Among all its modules, we will focus on `pyplot` that provides a comfortable interface with the plotting libraries. For example, if we desire to plot a cycle of the sine function, we could execute the following code snippet:

```
>>> import numpy
>>> import matplotlib.pyplot as plt
>>> x=numpy.linspace(0,2*numpy.pi,32)
>>> fig = plt.figure()
>>> plt.plot(x, numpy.sin(x))
>>> plt.show()
>>> fig.savefig('sine.png')
```

We obtain the following plot:

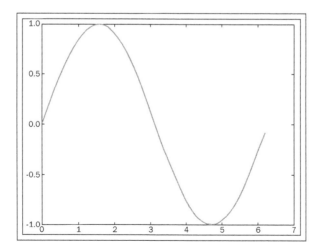

Let us explain each command from the previous session. The first two commands are used to import `numpy` and `matplotlib.pyplot` as usual. We define an array x of 32 uniformly spaced floating point values from 0 to 2π, and define y to be the array containing the sine of the values from x. The command figure creates space in the memory to store the subsequent plots and puts in place an object of the `matplotlib.figure.Figure` form. The `plt.plot(x, numpy.sin(x))` command creates an object of the `matplotlib.lines.Line2D` form containing data with the plot of x against `numpy.sin(x)` together with a set of axes attached to it and labeled according to the ranges of the variables. This object is stored in the previous `Figure` object and is displayed on the screen via the `plt.show()` command. The last command in the session, `fig.savefig()`, saves the Figure object to whatever valid image format we desire (in this case, a **Portable Network Graphics (PNG)** image). From now on, in any code that deals with matplotlib commands, we will leave the option of showing/saving open.

There are, of course, commands that control the style of axes, aspect ratio between axes, labeling, colors, legends, the possibility of managing several figures at the same time (subplots), and many more features to display all sorts of data. We will be discovering these as we progress with examples throughout the book.

How to open IPython Notebooks

This book comes with a set of IPython Notebooks that will help you interactively test and modify or adapt to your needs to the code snippets shown in each chapter of the book. We should warn, however, that these IPython Notebooks will make sense only if read along side the book.

In this regard, this book assumes familiarity with Python and some of its development environment as the IPython Notebook. Consequently, we will only refer to the documentation on the official website for IPython Notebook (`http://ipython.org/notebook.html`). You can find additional help at (`http://ipython.org/ipython-doc/stable/notebook/index.html`). Note that IPython Notebook is also available through **Wakari** (`https://wakari.io/`), as a standalone or part of the Anaconda package, or by Enthought. If you're new to IPython Notebook, get started by looking at the example collection and reading the documentation.

To use the files for this book, open a terminal and go to the directory where the file you want to open is stored (it should have the form `filename.ipynb`). At the command line, in that terminal, type:

```
ipython notebook filename.ipynb
```

After hitting the *enter* key, the file should be displayed in the default web browser. In case that does not happen, please note that the IPython Notebook is officially supported on the browsers Chrome, Safari, and Firefox. For additional details refers to the *Browser Compatibility* section on the documentation currently at `http://ipython.org/ipython-doc/stable/install/install.html`.

Once the `.ipynb` file has been opened, press and hold the *shift* key and hit *enter* to start executing the notebook cell by cell. Another way to execute the notebook cell by cell is via the player icon on the menu near the left of the cell labeled as **markdown**. Alternatively, from the **Cell** menu (on the top of the browser) you could choose among several options to execute the contents of the notebook.

To leave the notebook you could choose **Close and halt**, from the **File** menu on top of the browser below the label **Notebook**. Options to save the notebook can also be found under the **File** menu. To completely close the notebook browser you need to hit the keys *ctrl* and *C* simultaneously on the terminal where the notebook was started and follow the instructions after that.

Summary

In this chapter, you have learned the benefits of using the combination of Python, NumPy, SciPy, and matplotlib as a programming environment for any scientific endeavor that requires mathematics; in particular, anything related to numerical computations. You have explored the environment, learned how to download, install, and test the required libraries, used them for some quick computations, and figured out a few good ways to search for help.

In *Chapter 2*, *Working with the NumPy Array As a First Step to SciPy*, we will guide you through basic object creation in SciPy, including the best methods to manipulate data, or obtain information from it.

2
Working with the NumPy Array As a First Step to SciPy

At the top level, SciPy is basically NumPy, since both the object creation and basic manipulation of these objects are performed by functions of the latter library. This assures much faster computations, since the memory handling is done internally in an optimal way. For instance, if an operation must be made on the elements of a big multidimensional array, a novice user might be tempted to go over columns and rows with as many for loops as necessary. Loops run much faster when they access each consecutive element in the same order in which they are stored in memory. We should not be bothered with considerations of this kind when coding. The NumPy/SciPy operations assure that this is the case. As an added advantage, the names of operations in NumPy/SciPy are intuitive and self explanatory. Code written in this fashion is extremely easy to understand and maintain, faster to correct or change in case of need.

Let's illustrate this point with an introductory example.

The `scipy.misc` module in the SciPy package contains a classical image called `lena`, used in the image processing community for testing and comparison purposes. This is a 512 x 512 pixel standard test image, which has been in use since 1973, and was originally cropped from the centerfold of the November 1972 issue of the Playboy magazine. It is a picture of Lena Söderberg, a Swedish model, shot by photographer Dwight Hooker. The image is probably the most widely used test image for all sorts of image processing algorithms (such as compression and noise reduction) and related scientific publications.

This image is stored as a two-dimensional array. Note that the number in the nth column and m^{th} row of this array measures the grayscale value at the pixel position $(n+1, m+1)$ of the image. In the following, we access this picture and store it in the `img` variable, by issuing the following commands:

```
>>> import scipy.misc
>>> img=scipy.misc.lena()
>>> import matplotlib.pyplot as plt
>>> plt.gray()
>>> plt.imshow(img)
```

The image can be displayed by issuing the following command:

```
>>> plt.show()
```

We may take a peek at some of these values; say the 7 x 3 upper corner of the image (7 columns, 3 rows). Instead of issuing for loops, we could *slice* the corresponding portion of the image. The img[0:3,0:7] command gives us the following:

```
array([[162, 162, 162, 161, 162, 157, 163],
       [162, 162, 162, 161, 162, 157, 163],
       [162, 162, 162, 161, 162, 157, 163]])
```

We can use the same strategy to populate arrays or change their values. For instance, let's change all entries of the previous array to hold zeros on the second row between columns 2 to 6:

```
>>> img[1,1:6]=0
>>> print (img[0:3,0:7])
```

The output is shown as follows:

```
[[162 162 162 161 162 157 163]
 [162   0   0   0   0   0 163]
 [162 162 162 161 162 157 163]]
```

Object essentials

We have been introduced to NumPy's main object—the homogeneous multidimensional array, also referred to as `ndarray`. All elements of the array are casted to the same datatype (homogeneous). We obtain the datatype by the `dtype` attribute, its dimension by the `shape` attribute, the total number of elements in the array by the `size` attribute, and elements by referring to their positions:

```
>>> img.dtype, img.shape, img.size
```

The output is shown as follows:

```
(dtype('int64'), (512, 512), 262144)
```

Let's compute the grayscale values now:

```
>>> img[32,67]
```

The output is shown as follows:

```
87
```

Let's interpret the outputs. The elements of `img` are 64-bit integer values ('int64'). This may vary depending on the system, the Python installation, and the computer specifications. The shape of the array (note it comes as a Python tuple) is 512 x 512, and the number of elements 262144. The grayscale value of the image in the 33rd column and 68th row is `87` (note that in NumPy, as in Python or C, all indices are zero-based).

We will now introduce the basic property and methods of NumPy/SciPy objects—datatype and indexing.

Using datatypes

There are several approaches to impose the datatype. For instance, if we want all entries of an already created array to be 32-bit floating point values, we may cast it as follows:

```
>>> import scipy.misc
>>> img=scipy.misc.lena().astype('float32')
```

We can also use an optional argument, `dtype` through the command:

```
>>> import numpy
>>> scores = numpy.array([101,103,84], dtype='float32')
>>> scores
```

The output is shown as follows:

```
array([ 101.,   103.,    84.], dtype=float32)
```

This can be simplified even further with a third clever method (although this practice offers code that are not so easy to interpret):

```
>>> scores = numpy.float32([101,103,84])
>>> scores
```

The output is shown as follows:

```
array([ 101.,   103.,    84.], dtype=float32)
```

The choice of datatypes for NumPy arrays is very flexible; we may choose the basic Python types (including `bool`, `dict`, `list`, `set`, `tuple`, `str`, and `unicode`), although for numerical computations we focus on `int`, `float`, `long`, and `complex`.

NumPy has its own set of datatypes optimized to use with instances of `ndarray`, and with the same precision as the previously given native types. We distinguish them with a trailing underscore (_). For instance, `ndarray` of strings could be initialized, as follows:

```
>>> a=numpy.array(['Cleese', 'Idle', 'Gilliam'], dtype='str_')
>>> a.dtype
```

The output is shown as follows (it depends on your Python version):

```
dtype('<U7')
```

Note two things; unlike it's purely Python counterpart, the usage of the `'str_'` datatype requires the name to be quoted; we could use the longer unquoted version, `numpy.str_`.

When prompted for datatype, the system returns its C-derived equivalent: `'<U7'` (`'<U` for strings, and `7'` to indicate the largest size of any of its elements).

The most common way to address numerical types is with the bit width nomenclature: `boolXX`, `intXX`, `uintXX`, `floatXX`, or `complexXX`, where `XX` indicates the bit size (for example, `uint32` for 32-bit unsigned integers).

It is also possible to design our own datatypes, and this is where the full potential of the flexibility of NumPy datatypes arise. For instance, a datatype to indicate the name and grades of a student could be created, as follows:

```
>>> dt = numpy.dtype([ ('name', numpy.str_, 16), ('grades',
    numpy.float64, (2,)) ])
>>> dt
```

The output is shown as follows (it depends on your Python version):

```
dtype([('name', '<U16'), ('grades', '<f8', (2,))])
```

This means that the `dt` datatype has two parts: the first part, the `name`, that must be a `numpy.str_` string with 16 characters. The second part, the `grades`, is a subarray of dimension 2 with scores as 64-bit floating point values. A valid array with elements in this datatype would then look like the following:

```
>>> MA141=numpy.array([ ('Cleese', (7.0,8.0)), ('Gilliam',
    (9.0,10.0)) ], dtype=dt)
>>> MA141
```

The output is shown as follows (it depends on your Python version):

```
array([('Cleese', [7.0, 8.0]), ('Gilliam', [9.0, 10.0])],
    dtype=[('name', '<U16'), ('grades', '<f8', (2,))])
```

Indexing and slicing arrays

There are two basic methods to access the data in a NumPy array; let's call that array for A. Both methods use the same syntax, `A[obj]`, where `obj` is a Python object that performs the selection. We are already familiar with the first method of accessing a single element. The second method is the subject of this section, namely **slicing**. This concept is exactly what makes NumPy and SciPy so incredibly easy to manage.

The basic slice method is a Python object of the form `slice(start,stop,step)`, or in a more compact notation, `start:stop:step`. Initially, the three variables, `start`, `stop`, and `step` are non-negative integer values, with `start` less than or equal to `stop`.

This represents the sequence of indices k = *start + (i * step)*, where *k* runs from `start` to the largest integer *k_max = start + step*int((stop-start)/step)*, or *i* from 0 to the largest integer equal to *int((stop - start) / step)*. When a slice method is invoked on any of the dimensions of `ndarray`, it selects all elements in that dimension indexed by the corresponding sequence of indices. The simple example next illustrates this point:

```
>>> A=numpy.array([[1,2,3,4,5,6,7,8],[2,4,6,8,10,12,14,16]])
>>> print (A[0:2, 0:8:2])
```

The output is shown as follows:

```
[[ 1  3  5  7]
 [ 2  6 10 14]]
```

If `start` is greater than `stop`, a negative value of `step` is used to traverse the sequence backwards:

```
>>> print (A[0:2, 8:0:-2])
```

The output is shown as follows:

```
[[ 8,  6,  4,  2]
 [16, 12,  8,  4]]
```

Negative values of `start` and `stop` are interpreted as n-start and n-stop (respectively), where n is the size of the corresponding dimension. The A[0:2, -1:0:-2] command gives exactly the same output as the previous example.

The slice objects can be shortened by the absence of `start` (which implies a zero if `step` is positive, or the size of the dimension if `step` is negative), absence of `stop` (which implies the size of the corresponding dimension in case of positive `step`, or zero in case of negative `step`). Absence of `step` implies `step` is equal to 1. The `::` object can be shortened simply as `:` for an easier syntax. The A[:,::-2] command then offers, yet again, the same output as the previous two.

The first nonbasic method of accessing data from an array is based on the idea of collecting several indices and requesting the elements in the array with those indices. For example, from our previous array A, we would like to construct a new array with the elements on locations (0, 0), (0, 3), (1, 2), and (1, 5). We do so by gathering the *x* and *y* values of the indices in respective lists, [0,0,1,1] and [0,3,2,5], and feeding these lists to A as an indexing object, as follows:

```
>>> print (A[ [0,0,1,1], [0,3,2,5] ])
```

The output is shown as follows:

```
[ 1  4  6 12]
```

Note how the result loses the dimension of the primitive array, and offers a one-dimensional array. If we desire to capture a subarray of A with indices in the **Cartesian** product of two sets of indices, respecting the row and column choice and creating a new array with the dimensions of the Cartesian product, we use the `ix_` command. For instance, if in our previous array we would like to obtain the subarray of dimension 2 x 2 with indices in the Cartesian product of indices (0, 1) by (0,3) (these are the locations (0, 0), (0, 3), (1, 0), and (1, 3)), we do so as follows:

```
>>> print (A[ numpy.ix_ ( [0,1], [0,3] )])
```

The output is shown as follows:

```
[[1 4]
 [2 8]]
```

The array object

At this point, we are ready for a thorough study of all interesting attributes of ndarray for scientific computing purposes. We have already covered a few, such as dtype, shape, and size. Other useful attributes are ndim (to compute the number of dimensions in the array), real, and imag (to obtain the real and imaginary parts of the data, should this be formed by complex numbers) or flat (which creates a one-dimensional indexable iterator from the data).

For instance, if we desired to add all the values of an array together, we could use the flat attribute to run over all the elements sequentially, and accumulate all the values in a variable. A possible code to perform this task should look like the following code snippet (compare this code with the ndarray.sum() method, which will be explained in object calculation ahead):

```
>>> value=0; import scipy.misc; img=scipy.misc.lena()
>>> for item in img.flat: value+=item
>>> value
```

The output is shown as follows:

```
32518120
```

We will also explore some of the methods applied to arrays. These are tools used to modify objects; let it be their datatypes, their shape, or their structure through conversion. These methods can be classified in three big categories—**array conversion**, **shape selection/manipulation**, and **object calculation**.

Array conversions

The `astype()` method returns a copy of the array converted to a specific type; the `copy` method returns a copy of the array. Finally, the `tofile()`, `tolist()`, or `tostring()` method writes the binary data of the array into a file, returns a hierarchical python list version of the same array, or returns a string representation of the array data.

For instance, to write the contents of the `img` array to a text file making sure that each entry of the array is printed as an integer and that every two integers are separated by a white space, we can issue the following command:

```
>>> img.tofile("lena.txt",sep=" ",format="%i")
```

Note how the formatting string follows the C language conventions.

Shape selection/manipulations

These are used not only when we need to rearrange (`swapaxes` and `transpose`) or sort (`argsort` and `sort`) an array, but also when we need to reshape (`reshape`), resize (`flatten`, `ravel`, `resize`, and `squeeze`), or select (`choose`, `compress`, `diagonal`, `nonzero`, `searchsorted`, and `take`) arrays. Note that these methods are very powerful when combined with slicing operations; as a matter of fact, many of them can replace slicing to offer more readability.

We need to say a word about the attributes `flat`, `ravel`, and `flatten`, which offer very similar outputs, but very different memory management. The first attribute, `flat`, creates an iterator over an array. Once used, it disappears from memory. The attribute `ravel` returns a one-dimensional flattened array of the input; a copy is made only if needed. Finally, `flatten` creates a one-dimensional array of the input, and always allocates memory for it. We use it only when we need to change the values of flattened arrays. We will highlight the power of the sorting methods in the following code snippets. When sorting an array of integers, what would be the order of their indices? We may obtain this information with the `argsort()` method. We may even impose which sorting algorithm is to be used (rather than coding it ourselves)—`quicksort`, `mergesort`, or `heapsort`. We can even sort the array in place, using the `sort()` method. Let's take a look at the following set of commands:

```
>>> import numpy
>>> A = numpy.array([11,13,15,17,19,18,16,14,12,10])
>>> A.argsort(kind='mergesort')
```

The output is shown as follows:

```
array([9, 0, 8, 1, 7, 2, 6, 3, 5, 4])
```

Now, we apply the `sort()` method:

```
>>> A.sort()
>>> print(A)
```

The output is shown as follows:

```
[10 11 12 13 14 15 16 17 18 19]
```

Object calculations

Array calculation methods are used to perform computations or extract information from our data. Python supplies a range of statistical methods to compute, for instance, maximum and minimum values of the data (`max` and `min`) with their corresponding indices (`argmax` and `argmin`) methods to compute the sum, cumulative sums, product, or cumulative products (`sum`, `cumsum`, `prod`, and `cumprod`), and to calculate the average (`mean`), point spread (`ptp`), variance (`var`), and standard deviation (`std`) of our data. Other methods allow us to compute complex conjugate of complex-valued arrays (`conj`), the trace of the array (`trace`, which is the sum of the elements in the diagonal), and even clipping the matrix (`clip`) by forcing a minimum and maximum value below and above certain thresholds.

Note, that most of these methods can act on the entire array and each of their dimension:

```
>>> A=numpy.array([[1,1,1],[2,2,2],[3,3,3]])
>>> A.mean()
```

The output is shown as follows:

```
2
```

Now, let's apply the `mean()` method with `axis=0`:

```
>>> A.mean(axis=0)
```

The output is shown as follows:

```
array([ 2.,   2.,   2.])
```

Similarly, we perform the same command with `axis=1`:

```
>>> A.mean(axis=1)
```

The output is shown as:

```
array([ 1.,   2.,   3.])
```

Let's also illustrate the `clip` command with an easy exercise based on the Lena image. Compute the maximum and minimum values of Lena (`img`), and contrast them with the point spread (it should be equal to the difference between those two values). Now, create a new array `A` by clipping Lena so that the minimum is maintained, but the point spread is reduced to only 100 values. Let's illustrate the effect of `min()`, `max()`, and `ptp()` commands on Lena (`img`):

```
>>> img.min(), img.max(), img.ptp()
```

The output is shown as follows:

```
(25, 245, 220)
```

Further, we illustrate the effect of `clip()` command on `img` in the following lines of code:

```
>>> A=img.clip(img.min(),img.min()+100)
>>> A.min(), A.max(), A.ptp()
```

The output is shown as follows:

```
(25, 125, 100)
```

Array routines

In this section, we will deal with most operations on arrays. We will classify them into four main categories:

- Routines to create new arrays
- Routines to manipulate a single array
- Routines to combine two or more arrays
- Routines to extract information from arrays

The reader will surely realize that some operations of this kind can be carried out by methods, which once again shows the flexibility of Python and NumPy.

Routines to create arrays

We have previously seen the command to create an array and store it to a variable `A`. Let's take a look at it again:

```
>>> A=numpy.array([[1,2],[2,1]])
```

The complete syntax, however, writes as follows:

```
array(object,dtype=None,copy=True,order=None, subok=False,ndim=0)
```

Let's go over the options: object is simply the data we use to initialize the array. In the previous example, the object is a 2 x 2 square matrix; we may impose a datatype with the dtype option. The result is stored in the variable A. If copy is True, the returned object will be a copy of the array, if False, the returned object will only be a copy, if dtype is different from the datatype of object. The arrays are stored following a C-style ordering of rows and columns. If the user prefers to store the array following the memory style of FORTRAN, the order='Fortran' option should be used. The subok option is very subtle; if True, the array may be passed as a subclass of the object, if False, then only ndarray arrays are passed. And finally, the ndmin option indicates the smallest dimension returned by the array. If not offered, this is computed from object.

A set of special arrays can be obtained with commands such as zeros, ones, empty, identity, and eye. The names of these commands are quite informative:

- zeros creates an array filled with zeros.
- ones creates an array filled with ones.
- empty returns an array of required shape without initializing its entries.
- identity creates a square matrix with dimensions indicated by a single positive integer n. The entries are filled with zeros, except the diagonal, which is filled with ones.

The eye command is very similar to identity. It also constructs diagonal arrays, but unlike identity, eye allows specifying diagonals offset the traditional centered, as it can operate on rectangular arrays as well. In the following lines of code, we use zeros, ones, and identity commands:

```
>>> Z=numpy.zeros((5,5), dtype=int)
>>> U=numpy.ones((2,2), dtype=int)
>>> I=numpy.identity(3, dtype=int)
```

In the first two cases, we indicated the shape of the array (as a Python tuple of positive integers) and the optional datatype imposition.

The syntax for eye is as follows:

```
numpy.eye(N,M=None,k=0,dtype=float)
```

The integers, N and M indicate the shape of the array, and the integer k indicates the index of the diagonal to populate.

An index `k=0` (the default) points to the traditional diagonal; a positive index refers to upper diagonals and negative to lower diagonals. To illustrate this point, the following example shows how to create a 4 x 4 sparse matrix with nonzero elements on the first upper and subdiagonals:

```
>>> D=numpy.eye(4,k=1) + numpy.eye(4,k=-1)
>>> print (D)
```

The output is shown as follows:

```
[[ 0.  1.  0.  0.]
 [ 1.  0.  1.  0.]
 [ 0.  1.  0.  1.]
 [ 0.  0.  1.  0.]]
```

Using the previous four commands together with basic slicing, it is possible to create even more complex arrays very simply. We propose the following challenge.

Use exclusively, the previous definitions of U and I together with an `eye` array. How would the reader create a 5 x 5 array A of values, type float with *fives* at the four entries (0, 0), (0, 1), (1, 0), and (1, 1); *sixes* along the remaining entries of the diagonal; and *threes* in the two other corners ? The solution to this question can be addressed by issuing the following set of commands:

```
>>> A=3.0*(numpy.eye(5,k=4) + numpy.eye(5,k=-4))
>>> A[0:2,0:2]=5*U; A[2:5,2:5]=6*I
>>> print (A)
```

The output is shown as follows:

```
[[ 5.  5.  0.  0.  3.]
 [ 5.  5.  0.  0.  0.]
 [ 0.  0.  6.  0.  0.]
 [ 0.  0.  0.  6.  0.]
 [ 3.  0.  0.  0.  6.]]
```

The flexibility of creating an array in NumPy is even more clear using the `fromfunction` command. For instance, if we require a 4 x 4 array where each entry reflects the product of its indices, we may use the `lambda` function (`lambda i,j: i*j`) in the `fromfunction` command, as follows:

```
>>> B=numpy.fromfunction( (lambda i,j: i*j), (4,4), dtype=int)
>>> print (B)
```

The output is shown as follows:

```
[[0 0 0 0]
 [0 1 2 3]
 [0 2 4 6]
 [0 3 6 9]]
```

A very important tool dealing with arrays is the concept of masking. **Masking** is based on the idea of selecting or masking those indices for which their corresponding entries satisfy a given condition. For example, in the array B shown in the previous example, we can mask all zero-valued entries with the B==0 command, as follows:

```
>>> print (B==0)
```

The output is shown as follows:

```
[[ True  True  True  True]
 [ True False False False]
 [ True False False False]
 [ True False False False]]
```

Now, how would the reader update B so that all zero's would be replaced by the sum of the squares of their corresponding indices?

Multiplying a mask by a second array of the same shape offers a new array in which each entry is either zero (if the corresponding entry in the mask is False), or the entry of the second array (if the corresponding entry in the mask is True):

```
>>> B += numpy.fromfunction((lambda i,j:i*i+j*j), (4,4))*(B==0)
>>> print (B)
```

The output is shown as follows:

```
[[0 1 4 9]
 [1 1 2 3]
 [4 2 4 6]
 [9 3 6 9]]
```

Note that we have created a new array filled with Boolean values as the size of the original array and in each step. This isn't a big deal in these toy examples, but when handling large datasets, allocating too much memory could seriously slow down our computations and exhaust the memory of our system. Among the commands to create arrays, there are two in particular putmask and where, which facilitate the management of resources internally, thus speeding up the process.

Note, for example, when we look for all odd-valued entries in B, the resulting mask has size of 16, although the interesting entries are only eight:

```
>>> print (B%2!=0)
```

The output is shown as follows:

```
[[False   True False   True]
 [ True   True False   True]
 [False False False False]
 [ True   True False   True]]
```

The numpy.where() command helps us gather those entries more efficiently. Let's take a look at the following command:

```
>>> numpy.where(B%2!=0)
```

The output is shown as follows:

```
(array([0, 0, 1, 1, 1, 3, 3, 3], dtype=int32),
 array([1, 3, 0, 1, 3, 0, 1, 3], dtype=int32))
```

If we desire to change those entries (all odd), to, say they are *squares plus one*, we can use the numpy.putmask() command instead, and better manage the memory at the same time. The following is a sample code for the numpy.putmask() command:

```
>>> numpy.putmask( B, B%2!=0, B**2+1)
>>> print (B)
```

The output is shown as follows:

```
[[ 0  2  4 82]
 [ 2  2  2 10]
 [ 4  2  4  6]
 [82 10  6 82]]
```

Note how the putmask procedure updates the values of B, without the explicit need to make a new assignment.

There are three additional commands that create arrays in the form of meshes. The arange and linspace commands create uniformly spaced values between two numbers. In arange, we specify the spacing between elements; in linspace, we specify the desired number of elements in the mesh. The logspace command creates uniformly spaced values in a logarithmic scale between the logarithms of two numbers to the base 10. The user could think of these outputs as the support of univariate functions.

The following is a sample code for the `numpy.arrange()` command:

```
>>> L1=numpy.arange(-1,1,0.3)
```

```
>>> print (L1)
```

The output for the preceding lines of code is shown as follows:

```
[-1.  -0.7 -0.4 -0.1  0.2  0.5  0.8]
```

The following is a sample code for the `numpy.linspace()` command:

```
>>> L2=numpy.linspace(-1,1,4)
>>> print (L2)
```

The output is shown as follows:

```
[-1.          -0.33333333  0.33333333  1.        ]
```

The following is an example for the `numpy.logspace()` command:

```
>>> L3= numpy.logspace(-1,1,4)
>>> print (L3)
```

The output for the preceding lines of code is shown as follows:

```
[  0.1          0.46415888  2.15443469  10.        ]
```

Finally, `meshgrid`, `mgrid`, and `ogrid` create two two-dimensional arrays of dimensions *n x m*, containing the elements of two given one-dimensional arrays of dimensions *n* and *m*. It accomplished this by repeating the values of each array as necessary. The user could think of these outputs as the support of functions of two variables.

The first of these routines, `meshgrid`, accepts only arrays as input. The other two routines, `mgrid` and `ogrid`, accept only indexing objects (for example, slices). The difference between these last two is a matter of memory allocation; while `mgrid` allocates full arrays with all the data, `ogrid` only creates enough sets so that the corresponding `mgrid` command could be obtained by a proper Cartesian product.

Let's take a look at the following `meshgrid` command:

```
>>> print (numpy.meshgrid(L2,L3))
```

The output is shown as follows:

```
(array([[-1.        , -0.33333333,  0.33333333,  1.        ],
        [-1.        , -0.33333333,  0.33333333,  1.        ],
        [-1.        , -0.33333333,  0.33333333,  1.        ],
        [-1.        , -0.33333333,  0.33333333,  1.        ]]), array([[
```

```
0.1        ,     0.1     ,     0.1      ,     0.1       ],
         [   0.46415888,    0.46415888,    0.46415888,    0.46415888],
         [   2.15443469,    2.15443469,    2.15443469,    2.15443469],
         [ 10.         ,   10.         ,   10.        ,    10.         ]]))
```

Let's take a look at the following `mgrid` command:

```
>>> print (numpy.mgrid[0:5,0:5])
```

The output is shown as follows:

```
[[[0 0 0 0 0]
  [1 1 1 1 1]
  [2 2 2 2 2]
  [3 3 3 3 3]
  [4 4 4 4 4]]

 [[0 1 2 3 4]
  [0 1 2 3 4]
  [0 1 2 3 4]
  [0 1 2 3 4]
  [0 1 2 3 4]]]
```

Let's take a look at the following `ogrid` command:

```
>>> print (numpy.ogrid[0:5,0:5])
```

The output is shown as follows:

```
[array([[0],
        [1],
        [2],
        [3],
        [4]]), array([[0, 1, 2, 3, 4]])]
```

We would like to finish the subsection on creations of arrays by showing one of the most useful routines for image processing and differential equations—the `tile` command. Its syntax is very simple, and is shown as follows:

```
tile(A, reps)
```

This routine presents a very effective method of tiling an array A following some repetition pattern reps (a tuple, a list, or another array) to create larger arrays. The following checkerboards exercise shows its potential.

Start with two small binary arrays—B=numpy.ones((3,3)) and checker2by2=numpy.zeros((6,6)) and create a checkerboard using tile and as few operations as possible.

Let's perform some operations using these commands:

```
>>> checker2by2[0:3,0:3]=checker2by2[3:6,3:6]=B
>>> numpy.tile(checker2by2,(4,4))
```

The output is too long to be shown here. Please refer to the *How to open IPython Notebooks* section in *Chapter 1, Introduction to SciPy*, to run the IPython Notebook corresponding to this chapter.

Routines for the combination of two or more arrays

On occasion, we need to combine the data of two or more arrays together to solve a specific problem. The core NumPy libraries contain extremely efficient routines to carry out these computations, and we urge the reader to get familiar with them. They are constructed with state-of-the-art algorithms, and they make sure that usage of memory is minimum and the complexity optimal. Most relevant are the routines that operate on arrays as if they were matrices. These include matrix products (outer, inner, dot, vdot, tensordot, cross, and kron), array correlations (correlate and convolve), array stacking (concatenate, vstack, hstack, column_stack, row_stack, and dstack), and array comparison (allclose).

If you are well-versed in linear algebra, you will surely enjoy the matrix products included in NumPy. We will postpone their usage and analysis until we cover the SciPy module on linear algebra in *Chapter 3, SciPy for Linear Algebra*.

An excellent use for correlation of arrays is basic pattern-matching. For instance, the image in the following example (the text array) contains an image of a paragraph extracted from the Wikipedia page about Don Quixote, while the second array, letterE, contains an image of the letter *e*, which is actually a subarray obtained from the text array and represents the pattern to be matched.

First, we load the text image and performs some preprocessing on it in order to bring the image to the right format (as close as possible to the grayscale approximation) to have better performance on this naive approach of pattern matching. We do this by executing the following lines of code in a Python console:

```
>>> import scipy.ndimage
>>> import numpy as np
>>> import matplotlib.pyplot as plt
```

```
>>> text = scipy.ndimage.imread('Chap_02_text_image.png')
>>> text = np.mean(text.astype(float)/255,-1)*2-1
```

Second, the pattern for the letter *e* is identified:

```
>>> letterE = text[37:53,275:291]
```

Next, a fraction of the maximum value of the correlation of both arrays offers the location of all the *e* letters contained in the array `text`:

```
>>> corr = scipy.ndimage.correlate(text,letterE)
>>> eLocations = (corr >= 0.95 * corr.max())
```

The positions in the image of the pattern found for x are as follows:

```
>>> CorrLocIndex = np.where(eLocations==True)
>>> x=CorrLocIndex[1]
>>> x
```

The output is shown as follows:

```
array([ 283,  514,  583,  681,  722,  881,  929, 1023,   64,  188,  452,
        504,  892,  921, 1059, 1087, 1102, 1133,  118,  547,  690, 1066,
       1110,  330,  363,  519,  671,  913,  951, 1119,  120,  292,  441,
        516,  557,  602,  649,  688,  717,  747,  783,  813,  988, 1016,
        250,  309,  505,  691,  769,  876,  904, 1057,  224,  289,  470,
        596,  626,  780, 1027,  112,  151,  203,  468,  596,  751,  817,
        867,  203,  273,  369,  560,  599,  888, 1111,  159,  221,  260,
        352,  427,  861,  901, 1034, 1146,  325,  506,  558])
```

The positions in the image of the found pattern for y are as follows:

```
>>> y=CorrLocIndex[0]
>>> y
```

The output is shown as follows:

```
array([ 45,  45,  45,  45,  45,  45,  45,  45,  74,  74,  74,  74,  74,
        74,  74,  74,  74,  74, 103, 103, 103, 103, 103, 132, 132, 132,
       132, 132, 132, 132, 161, 161, 161, 161, 161, 161, 161, 161, 161,
       161, 161, 161, 161, 161, 190, 190, 190, 190, 190, 190, 190, 190,
       219, 219, 219, 219, 219, 219, 219, 248, 248, 248, 248, 248, 248,
       248, 248, 277, 277, 277, 277, 277, 277, 277, 306, 306, 306, 306,
       306, 306, 306, 306, 306, 335, 335, 335])
```

There are 86 elements, which are in fact the total number of the occurrence of the letter *e* in the text image, as can be verified by counting them. Whether the matching has been done correctly can be verified graphically, superposing each pair (x, y) of the pattern on the text image, as follows:

```
>>> thefig=plt.figure()
>>> plt.subplot(211)
<matplotlib.axes._subplots.AxesSubplot object at 0x7fb9b2390110>
>>> plt.imshow(text, cmap=plt.cm.gray, interpolation='nearest')
<matplotlib.image.AxesImage object at 0x7fb9b1f29410>
>>> plt.axis('off')
```

The output for `plt.axis()` is shown as follows:

```
(-0.5, 1199.5, 359.5, -0.5)
```

Now, let's move further in the code:

```
>>> plt.subplot(212)
<matplotlib.axes._subplots.AxesSubplot object at 0x7fb9b1f29890>
>>> plt.imshow(text, cmap=plt.cm.gray, interpolation='nearest')
<matplotlib.image.AxesImage object at 0x7fb9b1637e10>
>>> plt.autoscale(False)
>>> plt.plot(x,y,'wo',markersize=10)
[<matplotlib.lines.Line2D object at 0x7fb9b1647290>]
>>> plt.axis('off')
```

The output for `plt.axis()` is shown as follows:

```
(-0.5, 1199.5, 359.5, -0.5)
```

Finally, in the following `show()` command, we display a figure that superposes each pair (x, y) of the pattern on the text image:

```
>>> plt.show()
```

This results in the following screenshot (the first image is the text and the next is the text where all occurrences of letter *e* have been crossed out):

Alonso Quijano, the protagonist of the novel, is a retired country gentleman nearing fifty years of age, living in an unnamed section of La Mancha with his niece and housekeeper. While mostly a rational man of sound reason, his reading of books of chivalry in excess has had a profound effect on him, leading to the distortion of his perception and the wavering of his mental faculties. In essence, he believes every word of these books of chivalry to be true though, for the most part, the content of these books is clearly fiction. Otherwise, his wits, in regards to everything other than chivaltry, are intact. He decides to out as a knight-errant in search of adventure. He dons an old suit of armour, renames himself "Don Quixote de la Mancha," and names his skinny horse "Rocinante". He designates a neighboring farm girl as his lady love, renaming her Dulcinea del Toboso, while she knows nothing about this.

Alonso Quijano, th● protagonist of th● nov●l, is a r●tir●d country g●ntl●man n●aring fifty y●ars of ag●, living in an unnam●d s●ction of La Mancha with his ni●c● and hous●k●●p●r. Whil● mostly a rational man of sound r●ason, his r●ading of books of chivalry in ●xc●ss has had a profound ●ff●ct on him, l●ading to th● distortion of his p●rc●ption and th● wav●ring of his m●ntal facultl●s. In ●ss●nc●, h● b●li●v●s ●v●ry word of th●s● books of chivalry to b● tru● though, for th● most part, th● cont●nt of th●s● books is cl●arly fiction. Oth●rwis●, his wits, in r●gards to ●v●rything oth●r than chivaltry, ar● intact. H● d●cid●s to out as a knight-●rrant in s●arch of adv●ntur●. H● dons an old suit of armour, r●nam●s hims●lf "Don Quixot● d● la Mancha," and nam●s his skinny hors● "Rocinant●". H● d●signat●s a n●ighboring farm girl as his lady lov●, r●naming h●r Dulcin●a del Toboso, whil● sh● knows nothing about this.

A few words about stacking operations; we have a basic concatenation routine, `concatenate`, which joins a sequence of arrays together along a pre-determined axis. Of course, all arrays in the sequence must have the same dimensions, otherwise it obviously doesn't work. The rest of the stack operations are syntactic sugar for special cases of `concatenate` — `vstack` to glue arrays vertically, `hstack` to glue arrays horizontally, `dstack` to glue arrays in the third dimension, and so on.

Another impressive set of routines are set operations. They allow the user to handle one-dimensional arrays as if they were sets and perform the Boolean operations of intersection (`intersect1d`), union (`union1d`), set difference (`setdiff1d`), and set exclusive or (`setxor1d`). The results of these set operations return sorted arrays. Note that it is also possible to test whether all the elements in one array belong to a second array (`in1d`).

Routines for array manipulation

There is a sequence of splitting routines, designed to break up arrays into smaller arrays, in any given dimension—`array_split`, `split` (both are basic splitting along the indicated axis), `hsplit` (horizontal split), `vsplit` (vertical split), and `dsplit` (in the third axis). Let's illustrate these with a simple example:

```
>>> import numpy
>>> B = numpy.ones((3,3))
>>> checker2by2 = numpy.zeros((6,6))
>>> checker2by2[0:3,0:3] = checker2by2[3:6,3:6] = B
>>> print(checker2by2)
```

The output is shown as follows:

```
[[ 1.  1.  1.  0.  0.  0.]
 [ 1.  1.  1.  0.  0.  0.]
 [ 1.  1.  1.  0.  0.  0.]
 [ 0.  0.  0.  1.  1.  1.]
 [ 0.  0.  0.  1.  1.  1.]
 [ 0.  0.  0.  1.  1.  1.]]
```

Now, let's perform the vertical split:

```
>>> numpy.vsplit(checker2by2,3)
```

The output is shown as follows:

```
[array([[ 1.,  1.,  1.,  0.,  0.,  0.],
        [ 1.,  1.,  1.,  0.,  0.,  0.]]),
 array([[ 1.,  1.,  1.,  0.,  0.,  0.],
        [ 0.,  0.,  0.,  1.,  1.,  1.]]),
 array([[ 0.,  0.,  0.,  1.,  1.,  1.],
        [ 0.,  0.,  0.,  1.,  1.,  1.]])]
```

Applying a Python function on an array *usually* means applying the function to each element of the array. Note how the NumPy function `sin` works on an array, for example:

```
>>> a=numpy.array([-numpy.pi, numpy.pi])
>>> print (numpy.vstack((a, numpy.sin(a))))
```

The output is shown as follows:

```
[[ -3.14159265e+00    3.14159265e+00]
 [ -1.22464680e-16    1.22464680e-16]]
```

Note that the `sin` function was computed on each element of the array.

This works provided the function has been properly vectorized (which is the case with `numpy.sin`). Notice the behavior with non-vectorized Python functions. Let's define such a function for computing, for each value of x, the maximum between x and 100 without using any routine from the NumPy libraries:

```
# function max100
>>> def max100(x):
            return(x)
```

If we try to apply this function to the preceding array, the system raises an error, as follows:

```
>>> max100(a)
```

The output is an error which is shown as:

```
ValueError: The truth value of an array with more than one element is
ambiguous. Use a.any() or a.all()
```

We need to explicitly indicate to the system when we desire to apply one of our functions to arrays, as well as scalars. We do that with the `vectorize` routine, as follows:

```
>>> numpy.vectorize(max100)(a)
```

The output is shown as follows:

```
array([100, 100])
```

For our benefit, the NumPy libraries provide a great deal of already-vectorized mathematical functions. Some examples are `round_`, `fix` (to round the elements of an array to a desired number of decimal places), and `angle` (to provide the angle of the elements of an array, provided they are complex numbers) and any basic trigonometric (`sin`, `cos`, `tan`, `sic`), exponential (`exp`, `exp2`, `sinh`, `cosh`), and logarithmic functions (`log`, `log10`, `log2`).

We also have mathematical functions that treat the array as an output of multidimensional functions, and offer relevant computations. Some useful examples are `diff` (to emulate differentiation along any specified dimension, by performing discrete differences), `gradient` (to compute the gradient of the corresponding function), or `cov` (for the covariance of the array).

Sorting the whole array according to the values of the first axis is also possible with the `msort` and `sort_complex` routines.

Routines to extract information from arrays

Most of the routines to extract information are statistical in nature, which include `average` (which acts exactly as the `mean` method), `median` (to compute the statistical median of the array on any of its dimensions, or the array as a whole), and computation of histograms (`histogram`, `histogram2d`, and `histogramdd`, depending on the dimensions of the array). The other important set of routines in this category deal with the concept of bins for arrays of dimension one. This is more easily explained by means of examples. Take the array `A=numpy.array([5,1,1,2,1,1,2,2,10,3,3,4,5])`, the `unique` command finds the unique values in the array and presents them as sorted:

```
>>> numpy.unique(A)
```

The output is shown as follows:

```
array([ 1, 2, 3, 4, 5, 10])
```

For arrays such as `A`, in which all the entries are nonnegative integers, we can visualize the array `A` as a sequence of eleven bins labeled with numbers from 0 to 10 (the maximum value in the array). Each bin with label n contains the number of n's in the array:

```
>>> numpy.bincount(A)
```

The output is shown as follows:

```
array([0, 4, 3, 2, 1, 2, 0, 0, 0, 0, 1])
```

For arrays where some of the elements are not numbers (nan), NumPy has a set of routines that mimic methods to extract information, but disregard the conflicting elements—`nanmax`, `nanmin`, `nanargmax`, `nanargmin`, `nansum`, and so on:

```
>>> A=numpy.fromfunction((lambda i,j: (i+1)*(-1)**(i*j)), (4,4))
>>> print (A)
```

The output is shown as follows:

```
[[ 1.  1.  1.  1.]
 [ 2. -2.  2. -2.]
 [ 3.  3.  3.  3.]
 [ 4. -4.  4. -4.]]
```

Let's see the effect of `log2` on array A:

```
>>> B=numpy.log2(A)
__main__:1: RuntimeWarning: invalid value encountered in log2
>>> print (B)
```

The output is shown as follows:

```
[[ 0.          0.          0.          0.        ]
 [ 1.               nan  1.               nan]
 [ 1.5849625   1.5849625   1.5849625   1.5849625]
 [ 2.               nan  2.               nan]]
```

Let's take a look at the sum and nansum commands in the following line of code:

```
>>> numpy.sum(B), numpy.nansum(B)
```

The output is shown as follows:

```
(nan, 12.339850002884624)
```

Summary

In this chapter, we have explored in depth the creation and basic manipulation of the object array used by SciPy, as an overview of the NumPy libraries. In particular, we have seen the principles of slicing and masking, which simplify the coding of algorithms to the point of transforming an otherwise unreadable sequence of loops and primitive commands into an intuitive and self-explanatory set of object calls and methods. You also learned that the nonbasic modules in NumPy are replicated as modules in SciPy itself. The chapter roughly followed the same structure as the official NumPy reference (which the reader can access at the SciPy pages http://docs.scipy.org/doc/numpy/reference/). There are other good sources that cover NumPy with rigor, and we refer you to any of that material for a more detailed coverage of this topic.

In the next five chapters, we will be accessing the commands that make SciPy a powerful tool in numerical computing. The structure of those chapters is basically a reflection of the different SciPy modules structured in an order that allows building applications on top of each other.

3
SciPy for Linear Algebra

In this chapter, we will continue exploring the different SciPy modules through meaningful examples. We will start with the treatment of matrices (whether normal or sparse) with the modules on Linear Algebra—`linalg` and `sparse`. Note that `linalg` expands on the NumPy module with the same name.

This discipline of mathematics studies vector spaces and linear mappings between them. Matrices represent objects in this field in such a way that any property of the underlying objects may be obtained by performing adequate operations on the representing matrices. In this chapter, we assume that you are familiar with at least the basics of linear algebra, in particular with the notion of matrix multiplication, finding the determinant and inverse of a matrix, as well as their immediate applications in **vector calculus**.

Accordingly, in this chapter, we will explore how vectors and matrices are handled in Numpy/SciPy, how to create them, how to program standard mathematical operations between them, and how to represent this on a functional form. Next, we will solve linear system of equations expressed in the matrix form involving dense or sparse matrices. The corresponding IPython Notebook will help you test the functionality of the modules involved and modify each illustrative example according to your specific needs.

Vector creation

As mentioned in *Chapter 2*, *Working with the NumPy Array As a First Step to SciPy*, SciPy depends on NumPy's main object's `ndarray` data structure. You can look at one-dimensional arrays as vectors and vice versa (oriented points in an n-dimensional space). Consequently, a vector can be created via Numpy as follows:

```
>>> import numpy
>>> vectorA = numpy.array([1,2,3,4,5,6,7])
```

```
>>> vectorA
```

The output is shown as follows:

```
array([1, 2, 3, 4, 5, 6, 7])
```

We can also use already defined arrays to create a new candidate. Some examples were presented in the previous chapter. Here we can reverse the already created vector and assign it to a new one:

```
>>> vectorB = vectorA[::-1].copy()
>>> vectorB
```

The output is shown as follows:

```
array([7, 6, 5, 4, 3, 2, 1])
```

Notice that in this example, we have to make a copy of the reverse of the elements of vectorA and assign it to vectorB. This way, by changing elements of vectorB, the elements of vectorA remain unchanged, as shown here:

```
>>> vectorB[0]=123
>>> vectorB
```

The output is shown as follows:

```
array([123,   6,   5,   4,   3,   2,   1])
```

Let's look at vectorA:

```
>>> vectorA
```

The output is shown as follows:

```
array([1, 2, 3, 4, 5, 6, 7])
```

Let's make a copy of vectorA by reversing its elements and assigning it to vectorB:

```
>>> vectorB = vectorA[::-1].copy()
>>> vectorB
```

The output is shown as follows:

```
array([7, 6, 5, 4, 3, 2, 1])
```

In the last code statement, we repeated the previous assignment to vectorB, bringing it back to its initial values taking the reverse of vectorA, once again.

Vector operations

In addition to being mathematical entities studied in linear algebra, Vectors are widely used in physics and engineering as a convenient way to represent physical quantities as **displacement**, **velocity**, **acceleration**, force, and so on. Accordingly, basic operations between vectors can be performed via Numpy/SciPy operations as follows:

Addition/subtraction

Addition/subtraction of vectors does not require any explicit loop to perform them. Let's take a look at addition of two vectors:

```
>>> vectorC = vectorA + vectorB
>>> vectorC
```

The output is shown as follows:

```
array([8, 8, 8, 8, 8, 8, 8])
```

Further, we perform subtraction on two vectors:

```
>>> vectorD = vectorB - vectorA
>>> vectorD
```

The output is shown as follows:

```
array([ 6,  4,  2,  0, -2, -4, -6])
```

Scalar/Dot product

Numpy has the built-in function dot to compute the scalar (dot) product between two vectors. We show you its use computing the dot product of vectorA and vectorB from the previous code snippet:

```
>>> dotProduct1 = numpy.dot(vectorA,vectorB)
>>> dotProduct1
```

The output is shown as follows:

```
84
```

Alternatively, to compute this product we could perform the element-wise product between the components of the vectors and then add the respective results. This is implemented in the following lines of code:

```
>>> dotProduct2 = (vectorA*vectorB).sum()
>>> dotProduct2
```

The output is shown as follows:

84

Cross/Vector product – on three-dimensional space vectors

First, two vectors in 3 dimensions are created before applying the built-in function from NumPy to compute the cross product between the vectors:

```
>>> vectorA = numpy.array([5, 6, 7])
>>> vectorB = numpy.array([7, 6, 5])
>>> crossProduct = numpy.cross(vectorA,vectorB)
>>> crossProduct
```

The output is shown as follows:

```
array([-12,   24, -12])
```

Further, we perform a `cross` operation of `vectorB` over `vectorA`:

```
>>> crossProduct = numpy.cross(vectorB,vectorA)
>>> crossProduct
```

The output is shown as follows:

```
array([ 12, -24,   12])
```

Notice that the last expression shows the expected result that `vectorA` cross `vectorB` is the negative of `vectorB` cross `vectorA`.

Creating a matrix

In SciPy, a matrix structure is given to any one- or two-dimensional `ndarray`, with either the `matrix` or `mat` command. The complete syntax is as follows:

```
numpy.matrix(data=object, dtype=None, copy=True)
```

Creating matrices, the data may be given as `ndarray`, a string or a Python list (as the second example below), which is very convenient. When using strings, the semicolon denotes change of row and the comma, change of column:

```
>>> A=numpy.matrix("1,2,3;4,5,6")
>>> A
```

The output is shown a follows s:

```
matrix([[1, 2, 3],
        [4, 5, 6]])
```

Let's look at another example:

```
>>> A=numpy.matrix([[1,2,3],[4,5,6]])
>>> A
```

The output is shown as follows:

```
matrix([[1, 2, 3],
        [4, 5, 6]])
```

Another technique to create a matrix from a two-dimensional array is to enforce the matrix structure on a new object, copying the data of the former with the `asmatrix` routine.

A matrix is said to be sparse (http://en.wikipedia.org/wiki/Sparse_matrix) if most of its entries are zeros. It is a waste of memory to input such matrices in the usual way, especially if the dimensions are large. SciPy provides different procedures to store such matrices effectively in memory. Most of the usual methods to input sparse matrices are contemplated in SciPy as routines in the `scipy.sparse` module. Some of those methods are **block sparse row** (`bsr_matrix`), **coordinate format** (`coo_matrix`), compressed sparse column or row (`csc_matrix`, `csr_matrix`), sparse matrix with diagonal storage (`dia_matrix`), dictionary with **Keys-based sorting** (`dok_matrix`), and **Row-based linked list** (`lil_matrix`).

At this point, we would like to present one of these: the coordinate format. In this format, and given a sparse matrix A, we identify the coordinates of the nonzero elements, say *n* of them, and we create two n-dimensional `ndarray` arrays containing the columns and the rows of those entries, and a third `ndarray` containing the values of the corresponding entries. For instance, notice the following sparse matrix:

$$\begin{pmatrix} 0 & 10 & 0 & 0 & 0 \\ 0 & 0 & 20 & 0 & 0 \\ 0 & 0 & 0 & 30 & 0 \\ 0 & 0 & 0 & 0 & 40 \\ 0 & 0 & 0 & 0 & 0 \end{pmatrix}$$

The standard form of creating such matrices is as follows:

```
>>> A=numpy.matrix([ [0,10,0,0,0], [0,0,20,0,0], [0,0,0,30,0],
                     [0,0,0,0,40], [0,0,0,0,0] ])
>>> A
```

The output is shown as follows:

```
matrix([[ 0, 10,  0,  0,  0],
        [ 0,  0, 20,  0,  0],
        [ 0,  0,  0, 30,  0],
        [ 0,  0,  0,  0, 40],
        [ 0,  0,  0,  0,  0]])
```

A more memory-efficient way to create these matrices would be to properly store the nonzero elements. In this case, one of the nonzero entries is at the 1st row and 2nd column (or location (0, 1) in Python) with value, 10. Another nonzero entry is at (1, 2) with value, 20. A 3rd nonzero entry, with the value 30, is located at (2, 3). The last nonzero entry of A is located at (3, 4), and has the value, 40.

We then have ndarray of rows, ndarray of columns, and another ndarray of values:

```
>>> import numpy
>>> rows=numpy.array([0,1,2,3])
>>> cols=numpy.array([1,2,3,4])
>>> vals=numpy.array([10,20,30,40])
```

We create the matrix A as follows:

```
>>> import scipy.sparse
>>> A=scipy.sparse.coo_matrix( (vals,(rows,cols)) )
>>> print (A); print (A.todense())
```

The output is shown as follows:

```
  (0, 1)    10
  (1, 2)    20
  (2, 3)    30
  (3, 4)    40
[[  0.  10.   0.   0.   0.]
 [  0.   0.  20.   0.   0.]
 [  0.   0.   0.  30.   0.]
 [  0.   0.   0.   0.  40.]]
```

Notice how the `todense` method turns sparse matrices into full matrices. Also note that it obviates any row or column of full zeros following the last nonzero element.

Associated to each input method, we have functions that identify sparse matrices of each kind. For instance, if we suspect that A is a sparse matrix in the `coo_matrix` format, we may use the following command:

```
>>> scipy.sparse.isspmatrix_coo(A)
```

The output is shown as follows:

```
True
```

All the array routines are cast to matrices, provided the input is a matrix. This is very convenient for matrix creation, especially thanks to stacking commands (`hstack`, `vstack`, `tile`). Besides these, matrices enjoy one more amazing stacking command, `bmat`. This routine allows the stacking of matrices by means of strings, making use of the convention: semicolon for change of row and comma for change of column. Also, it allows matrix names inside of the string to be evaluated. The following example is enlightening:

```
>>> B=numpy.mat(numpy.ones((3,3)))
>>> W=numpy.mat(numpy.zeros((3,3)))
>>> print (numpy.bmat('B,W;W,B'))
```

The output is shown as follows:

```
[[ 1.  1.  1.  0.  0.  0.]
 [ 1.  1.  1.  0.  0.  0.]
 [ 1.  1.  1.  0.  0.  0.]
 [ 0.  0.  0.  1.  1.  1.]
 [ 0.  0.  0.  1.  1.  1.]
 [ 0.  0.  0.  1.  1.  1.]]
```

The main difference between arrays and matrices is in regards to the behavior of the product of two objects of the same type. For example, multiplication between two arrays means *element-wise multiplication of the entries of the two arrays* and requires two objects of the same shape. The following code snippet is an example of multiplication between two arrays:

```
>>> a=numpy.array([[1,2],[3,4]])
>>> a*a
```

The output is shown as follows:

```
array([[ 1,  4],
       [ 9, 16]])
```

On the other hand, matrix multiplication requires a first matrix with shape (m, n), and a second matrix with shape (n, p) — the number of columns in the first matrix must be the same as the number of rows in the second matrix. This operation offers a new matrix of shape (m, p), as shown in the following diagram:

$$\begin{pmatrix} 1 & 2 \\ 3 & 4 \end{pmatrix} \cdot \begin{pmatrix} 1 & 2 \\ 3 & 4 \end{pmatrix} = \begin{pmatrix} 7 & 10 \\ 15 & 22 \end{pmatrix}$$

The following is the code snippet:

```
>>> A=numpy.mat(a)
>>> A*A
```

The output is shown as follows:

```
matrix([[ 7, 10],
        [15, 22]])
```

Alternatively, to obtain the matrix product between two conforming matrices as ndarray objects, we don't really need to transform the ndarray object to a matrix object if not needed. The matrix product could be obtained directly via the numpy. dot function introduced earlier in the *Scalar/Dot product* section of this chapter. Let's take a look at the following numpy.dot command example:

```
>>> b=numpy.array([[1,2,3],[3,4,5]])
>>> numpy.dot(a,b)
```

The output is shown as follows:

```
array([[ 7, 10, 13],
       [15, 22, 29]])
```

If we desire to perform an element-wise multiplication of the elements of two matrices, we can do so with the versatile numpy.multiply command, as follows:

```
>>> numpy.multiply(A,A)
```

The output is shown as follows:

```
matrix([[ 1,  4],
        [ 9, 16]])
```

The other difference between arrays and matrices worth noticing is in regard to their shapes. While we allow arrays to have one dimension; their corresponding matrices must have at least two. This is very important to have in mind when we transpose either object. Let's take a look at the following code snippet implementing `shape()` and `transpose()` commands:

```
>>> a=numpy.arange(5); A=numpy.mat(a)
>>> a.shape, A.shape, a.transpose().shape, A.transpose().shape
```

The output is shown as follows:

```
((5,), (1, 5), (5,), (5, 1))
```

As it has been shown, SciPy offers quite a number of basic tools to instantiate and manipulate matrices, with many related methods to follow. This also allows us to speed up computations in the cases where special matrices are used.

The `scipy.linalg` module provides commands to create special matrices such as block diagonal matrices from provided arrays (`block_diag`), **circulant matrices** (circulant), companion matrices (`companion`), **Hadamard matrices** (hadamard), **Hankel matrices** (hankel), **Hilbert** and **inverse Hilbert matrices** (hilbert, invhilbert), **Leslie matrices** (leslie), **square Pascal matrices** (pascal), **Toeplitz matrices** (toeplitz), **lower-triangular matrices** (tril), and **upper-triangular matrices** (triu).

Let's see an example on **optimal weighings**.

Suppose we are given p objects to be weighed in n weighings with a two-pan balance. We create an $n \times p$ matrix of plus and minus one, where a positive value in the (i, j) position indicates that the jth object is placed in the left pan of the balance in the ith weighing and a negative value that the jth object corresponding is in the right pan.

It is known that optimal weighings are designed by submatrices of Hadamard matrices. For the problem of designing an optimal weighing for eight objects with three weighings, we could then explore different choices of three rows of a Hadamard matrix of order eight. The only requirement is that the sum of the elements on the row of the matrix is zero (so that the same number of objects are placed on each pan). Through slicing, we can accomplish just that:

```
>>> import scipy.linalg
>>> A=scipy.linalg.hadamard(8)
>>> zero_sum_rows = (numpy.sum(A,0)==0)
>>> B=A[zero_sum_rows,:]
>>> print (B[0:3,:])
```

The output is shown as follows:

```
[[ 1 -1  1 -1  1 -1  1 -1]
 [ 1  1 -1 -1  1  1 -1 -1]
 [ 1 -1 -1  1  1 -1 -1  1]]
```

The `scipy.sparse` module has its own set of special matrices. The most common are matrices of those along diagonals (`eye`), identity matrices (`identity`), matrices from diagonals (`diags`, `spdiags`), block diagonal matrices from sparse matrices (`block_diag`), matrices from sparse sub-blocks (`bmat`), column-wise and row-wise stacks (`hstack`, `vstack`), and random matrices of a given shape and density with uniformly distributed values (`rand`).

Matrix methods

Besides inheriting all the array methods, matrices enjoy four extra attributes: `T` for transpose, `H` for conjugate transpose, `I` for inverse, and `A` to cast as `ndarray`:

```
>>> A = numpy.matrix("1+1j, 2-1j; 3-1j, 4+1j")
>>> print (A.T); print (A.H)
```

The output is shown as follows:

```
[[ 1.+1.j  3.-1.j]
 [ 2.-1.j  4.+1.j]]
[[ 1.-1.j  3.+1.j]
 [ 2.+1.j  4.-1.j]]
```

Operations between matrices

We have briefly covered the most basic operation between two matrices; the matrix product. For any other kind of product, we resort to the basic utilities in the NumPy libraries, as: dot product for arrays or vectors (`dot`, `vdot`), inner and outer products of two arrays (`inner`, `outer`), **tensor dot product** along specified axes (`tensordot`), or the **Kronecker product** of two arrays (`kron`).

Let's see an example of creating an **orthonormal** basis.

Create an orthonormal basis in the nine-dimensional real space from an orthonormal basis of the three-dimensional real space.

Let's choose, for example, the orthonormal basis formed by the vectors as shown in following diagram:

$$
\begin{aligned}
v_1 &= \frac{1}{\sqrt{2}}(1,0,1), \\
v_2 &= (0,1,0), \\
v_3 &= \frac{1}{\sqrt{2}}(1,0,-1)
\end{aligned}
$$

We compute the desired basis by collecting these vectors in a matrix and using a Kronecker product, as follows:

```
>>> import numpy
>>> import scipy.linalg
>>> mu = 1/numpy.sqrt(2)
>>> A = numpy.matrix([[mu,0,mu],[0,1,0],[mu,0,-mu]])
>>> B = scipy.linalg.kron(A,A)
```

The columns of matrix B shown previously, give us an orthonormal basis directly. For instance, the vectors with odd indices would be the columns of the following submatrix:

```
>>> print (B[:,0:-1:2])
```

The output is shown as follows:

```
[[ 0.5   0.5  0.    0.5]
 [ 0.    0.    0.    0. ]
 [ 0.5  -0.5  0.    0.5]
 [ 0.    0.    0.    0. ]
 [ 0.    0.    1.    0. ]
 [ 0.   -0.   0.    0. ]
 [ 0.5   0.5  0.   -0.5]
 [ 0.    0.    0.   -0. ]
 [ 0.5  -0.5  0.   -0.5]]
```

Functions on matrices

The `scipy.linalg` module offers a useful set of functions on matrices. The basic two commands on square matrices are `inv` (for the inverse of a matrix) and `det` (for the determinant). The power of a square matrix is given by the standard exponentiation; that is, if A is a square matrix, then `A**2` indicates the matrix product `A*A`, which is shown in the following code snippet:

```
>>> A=numpy.matrix("1,1j;21,3")
>>> A; A*A; A**2
```

The output is shown as follows:

```
matrix([[  1.+0.j,    0.+1.j],
        [ 21.+0.j,    3.+0.j]])
matrix([[  1.+21.j,   0. +4.j],
        [ 84. +0.j,   9.+21.j]])
matrix([[  1.+21.j,   0. +4.j],
        [ 84. +0.j,   9.+21.j]])
```

It should be pointed out that as a type array, the product of `A*A` (or `A**2`) is calculated by squaring each element of the array:

```
>>> numpy.asarray(A); numpy.asarray(A)*numpy.asarray(A);
    numpy.asarray(A)**2
```

The output is shown as follows:

```
array([[  1.+0.j,    0.+1.j],
       [ 21.+0.j,    3.+0.j]])
array([[   1.+0.j,   -1.+0.j],
       [ 441.+0.j,    9.+0.j]])
array([[   1.+0.j,   -1.+0.j],
       [ 441.+0.j,    9.+0.j]])
```

More advanced commands compute matrix functions that rely on the power series representation of expressions involving matrix powers, such as the matrix **exponential** (for which there are three possibilities — `expm`, `expm2`, and `expm3`), the matrix **logarithm** (`logm`), matrix **trigonometric functions** (`cosm`, `sinm`, `tanm`), matrix **hyperbolic trigonometric functions** (`coshm`, `sinhm`, `tanhm`), the **matrix sign function** (`signm`), or the matrix **square root** (`sqrtm`).

Notice the difference between the application of the normal exponential function on a matrix, and the result of a matrix exponential function.

In the former case, we obtain the application of numpy.exp to each entry of the matrix; in the latter, we actually compute the exponential of the matrix following the power series representation:

$$e^A = \sum_{n=0}^{\infty} \frac{1}{n!} A^n$$

The preceding formula is illustrated in this code snippet:

```
>>> import numpy
>>> import scipy.linalg
>>> a=numpy.arange(0,2*numpy.pi,1.6)
>>> A = scipy.linalg.toeplitz(a)
>>> print (A)
```

The output is shown as follows:

```
[[ 0.   1.6  3.2  4.8]
 [ 1.6  0.   1.6  3.2]
 [ 3.2  1.6  0.   1.6]
 [ 4.8  3.2  1.6  0. ]]
```

Let's perform the exp() operation on A:

```
>>> print (numpy.exp(A))
```

The output is shown as follows:

```
[[   1.          4.95303242   24.5325302   121.51041752]
 [   4.95303242  1.            4.95303242   24.5325302  ]
 [  24.5325302   4.95303242   1.            4.95303242]
 [ 121.51041752  24.5325302    4.95303242   1.          ]]
```

Let's perform the expm() operation on A:

```
>>> print (scipy.linalg.expm(A))
```

The output is shown as follows:

```
[[ 1271.76972856  916.49316549   916.63015271  1271.70874469]
 [  916.49316549  660.86560972   660.5306514    916.63015271]
 [  916.63015271  660.5306514    660.86560972   916.49316549]
 [ 1271.70874469  916.63015271   916.49316549  1271.76972856]]
```

For sparse square matrices, we have an optimized inverse function, as well as a matrix exponential—`scipy.sparse.linalg.inv`, `scipy.sparse.linalg.expm`.

For general matrices, we have the basic norm function (norm), as well as two versions of the **Moore-Penrose pseudoinverse** (`pinv` and `pinv2`).

Once again, we need to emphasize how important it is to rely on these functions, rather than coding their equivalent expressions manually. For instance, note the `norm` computation of vectors or matrices, `scipy.linalg.norm`. Let's show you, by example, the 2-norm of a two-dimensional vector `v=numpy.matrix([x,y])`, where at least one of the x and y values is extremely large—large enough so that x*x overflows:

```
>>> import numpy
>>> import scipy.linalg
>>> x=10**100; y=9; v=numpy.matrix([x,y])
>>> scipy.linalg.norm(v,2)
```

The output is shown as follows:

```
1e+100
```

Now, let's perform the `sqrt()` operation:

```
>>> numpy.sqrt(x*x+y*y)
```

The output is an error which is shown as follows:

```
Traceback (most recent call last)
  File "<stdin>", line 1, in <module>
AttributeError: 'long' object has no attribute 'sqrt'
```

Eigenvalue problems and matrix decompositions

Another set of operations heavily used on matrices is to compute and handle eigenvalues and eigenvectors of square matrices. These two problems rank among the most complex operations that we can perform on square matrices, and extensive research has been put in place to obtain good algorithms with low complexity and optimal usage of memory resources. SciPy has state-of-the-art code to implement these ideas.

For the computation of eigenvalues, the `scipy.linalg` module provides three routines: `eigvals` (for any ordinary or general eigenvalue problem), `eigvalsh` (if the matrix is symmetric of complex **Hermitian**), and `eigvals_banded` (if the matrix is banded). To compute the eigenvectors, we similarly have three corresponding choices: `eig`, `eigh`, and `eigh_banded`.

The syntax used in all cases is very similar. For example, for the general case of eigenvalues, we use the following line of code where matrix A must be square:

```
eigvals(A, B=None, overwrite_a=False)
```

This should be the only parameter passed to the routine if we wish to solve an ordinary eigenvalue problem. If we wish to generalize this, we may provide an extra square matrix (of the same dimensions as matrix A). This is passed in the B parameter.

The module also offers an extensive collection of functions that compute different decompositions of matrices, as follows:

- **Pivoted LU decomposition**: This function allows us to use the `lu` and `lufactor` commands.

- **Singular value decomposition**: This function allows us to use the `svd` command. To compute the singular values, we issue `svdvals`. If we wish to compose the sigma matrix in the singular value decomposition from its singular values, we do so with the `diagsvd` routine. If we wish to compute an orthogonal basis for the range of a matrix using SVD, we can accomplish this with the `orth` command.

- **Cholesky decomposition**: This function allows us to use the `cholesky`, `cholesky_banded`, and `cho_factor` commands.

- **QR and QZ decompositions**: This function allows us to use the `qr` and `qz` commands. If we wish to multiply a matrix with the matrix Q of a decomposition, we use the syntactic sugar `qr_multiply`, rather than performing this procedure in two steps.

- **Schur and Hessenberg decompositions**: This function allows us to use the `schur` and `Hessenberg` commands. If we wish to convert a real Schur form to complex, we have the `rsf2csf` routine.

At this point, we have an interesting application—image compression, which makes use of some of the routines explained so far.

Image compression via the singular value decomposition

This is a very simple application where a square image A of size n x n, and stored as ndarray is regarded as a matrix, and where a singular value decomposition (SVD) is performed on it. This operation is visible in the following diagram:

$$A = U \cdot S \cdot V^*, \ U = \begin{pmatrix} u_1 \\ \vdots \\ u_n \end{pmatrix}, S = \begin{pmatrix} s_1 & & \\ & \ddots & \\ & & s_n \end{pmatrix}, V^* = \begin{pmatrix} v_1 & \cdots & v_n \end{pmatrix}$$

From all the singular values of s we choose a fraction, together with their corresponding left and right singular vectors u, v. We compute a new matrix by collecting them according to the formula given in the following diagram:

$$\sum_{j=1}^{k} s_j \left(u_j \cdot v_j \right)$$

Note, for example, the similarity between the original (512 singular values) and an approximation using only 32 singular values:

```
>>> import numpy
>>> import scipy.misc
>>> from scipy.linalg import svd
>>> import matplotlib.pyplot as plt
>>> img=scipy.misc.lena()
>>> U,s,Vh=svd(img)        # Singular Value Decomposition
>>> A = numpy.dot( U[:,0:32],   # use only 32 singular values
        numpy.dot( numpy.diag(s[0:32]),
             Vh[0:32,:]))
>>> plt.subplot(121,aspect='equal'); plt.imshow(img); plt.gray()
>>> plt.subplot(122,aspect='equal'); plt.imshow(A)
>>> plt.show()
```

This produces the following images, of which the picture to the left is the original image and the picture to the right, the approximation using 32 singular values:

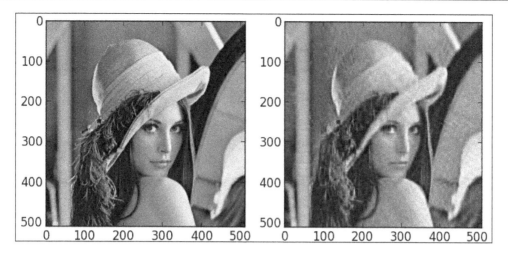

Using the `svd` approximation we managed to compress the original image of 262,144 coefficients (512 * 512)to only 32,800 coefficients ((2 * 32 * 512) + 32), or to one-eighth of the original information.

Solvers

One of the fundamental applications of linear algebra is to solve large systems of linear equations. For the basic systems of the form $Ax=b$, for any square matrix A and general matrix b (with as many rows as columns in A), we have two generic methods to find x (`solve` for dense matrices and `spsolve` for sparse matrices), using the following syntax:

```
solve(A, b, sym_pos=False, lower=False, overwrite_a=False,
overwrite_b=False, debug=False)
spsolve(A, b[, permc_spec, use_umfpack])
```

There are solvers that are even more sophisticated in SciPy, with enhanced performance for situations in which the structure of the matrix A is known. For dense matrices we have three commands in the `scipy.linalg` module: `solve_banded` (for banded matrices), `solveh_banded` (if besides banded, A is Hermitian), and `solve_triangular` (for triangular matrices).

When a solution is not possible (for example, if A is a singular matrix), it is still possible to obtain a matrix x that minimizes the `norm` of b-Ax in a least-squares sense. We can compute such a matrix with the `lstsq` command, which has the following syntax:

```
lstsq(A, b, cond=None, overwrite_a=False, overwrite_b=False)
```

The output of this function is a tuple that contains the following:

- The solution found (as ndarray)
- The sum of residues (as another ndarray)
- The effective rank of the matrix A
- The singular values of the matrix A (as another ndarray)

Let's illustrate this routine with a simple example, to solve the following system:

$$\begin{pmatrix} 0 & 1 & 0 \\ 0 & 0 & 1 \\ 0 & 0 & 0 \end{pmatrix} \cdot \begin{pmatrix} x \\ y \\ z \end{pmatrix} = \begin{pmatrix} 1 \\ 2 \\ 3 \end{pmatrix}$$

The following is the code snippet:

```
>>> import numpy
>>> import scipy.linalg
>>> A=numpy.mat(numpy.eye(3,k=1))
>>> print(A)
```

The output is shown as follows:

```
[[ 0.   1.   0.]
 [ 0.   0.   1.]
 [ 0.   0.   0.]]
```

Let's move further into the code and perform the following operations on b:

```
>>> b=numpy.mat(numpy.arange(3) + 1).T
>>> print(b)
```

The output is shown as follows:

```
[[1]
 [2]
 [3]]
```

Further, let's perform the lstsq operation:

```
>>> xinfo=scipy.linalg.lstsq(A,b)
>>> print (xinfo[0].T)        # output the solution
```

The output is shown as follows:

```
[[ 0.   1.   2.]]
```

The `overwrite_` options are designed to enhance performance of the algorithms, and should be used carefully, since they destroy the original data.

The truly fastest solvers in SciPy are based upon decomposition of matrices. Reducing the system into something simpler easily solves huge and really complicated systems of linear equations. We may accomplish this using decomposition techniques presented in the *Eigenvalue problems and matrix decompositions* and *Image compression via the singular value decomposition* subsections under the *Matrix methods* section of this chapter, but of course, the SciPy philosophy is to help us deal with all nuisances of memory and resources internally. To this end, the module also has the `lu_solve` (for solutions based on LU decompositions), and `cho_solve`, `cho_solve_banded` (for solutions based on Cholesky decompositions).

Finally, you will also find solvers for very complex matrix equations—the **Sylvester** equation (`solve_sylvester`), both the continuous and discrete algebraic **Riccati** equations (`solve_continuous_are`, `solve_discrete_are`) and both the continuous and discrete **Lyapunov** equations (`solve_discrete_lyapunov`, `solve_lyapunov`).

Most of the matrix decompositions and solutions to eigenvalue problems are contemplated for sparse matrices in the `scipy.sparse.linalg` module with a similar naming convention, but with much more robust use of computer resources and error control.

Summary

This chapter explored the treatment of vectors, matrices (whether normal or sparse) with the modules on linear algebra—`linalg` and `sparse.linalg`, which expand and improve the NumPy module with the same name.

In *Chapter 4, SciPy for Numerical Analysis*, we will continue discussing details of the options available in SciPy to perform numerical computations efficiently, will cover how to evaluate special functions found in applied mathematics and mathematical physics problems. This will be discussed in details of doing regression, interpolation and optimization via SciPy.

4
SciPy for Numerical Analysis

Practically all the different areas of numerical analysis are contemplated in some SciPy module. For example, in order to compute values of special functions, we use the `scipy.special` module. The `scipy.interpolate` module takes care of interpolation, extrapolation, and regression. For optimization, we have the `scipy.optimize` module, and finally, we have the `scipy.integrate` module for numerical evaluation of integrals. This last module serves as the interface to perform numerical solutions of ordinary differential equations as well.

Thus, in this chapter, we will first extensively explore how to use SciPy to numerically evaluate the special functions that are commonly found in the field of mathematical physics. Then, we will discuss the modules available in SciPy to tackle regression, interpolation, and optimization problems.

The chapter ends with a solution of the chaotic Lorenz system as an illustration of the capabilities included in SciPy to find numerical solutions of ordinary differential equations. The corresponding IPython Notebook will help you to try the functionalities of the modules involved in the computations and to modify each illustrative example according to your specific needs.

The evaluation of special functions

The `scipy.special` module contains numerically stable definitions of useful functions. Most often, the straightforward evaluation of a function at a single value is not very efficient. For instance, we would rather use a Horner scheme (http://en.wikipedia.org/wiki/Horner%27s_method) to find the value of a polynomial at a point than use the raw formula. The NumPy and SciPy modules ensure that this optimization is always guaranteed with the definition of all its functions, whether by means of Horner schemes or with more advanced techniques.

Convenience and test functions

All the convenience functions are designed to facilitate a computational environment where the user does not need to worry about relative errors. The functions seem to be pointless at first sight, but behind their codes, there are state-of-the-art ideas that offer faster and more reliable results.

We have convenience functions beyond the ones defined in the NumPy libraries to find the solutions of trigonometric functions in degrees (cosdg, sindg, tandg, and cotdg); to compute angles in radians from their expressions in degrees, minutes, and seconds (radian); common powers (exp2 for $2^{**}x$, and exp10 for $10^{**}x$); and common functions for small values of the variable (log1p for *log(1 + x)*, expm1 for *exp(x) - 1*, and cosm1 for *cos(x) - 1*).

For instance, in the following code snippet, the log1p function computes the natural logarithm of *1 + x*. Why not simply add 1 to the value of *x* and then take the logarithm instead? Let's compare:

```
>>> import numpy
>>> import scipy.special
>>> a=scipy.special.exp10(-16)
>>> numpy.log(1+a)
```

The output is as follows:

```
0.0
```

Now let's use log1p() on a:

```
>>> scipy.special.log1p(a)
```

The output is as follows:

```
9.9999999999999998e-17
```

While the absolute error of the first computation is small, the relative error is 100 percent.

In the same way as Lena image is regarded as the performance test in image processing, we have a few functions that are used to test different algorithms in different scenarios.

For instance, it is customary to test minimization codes against the Rosenbrock's banana function (http://en.wikipedia.org/wiki/Rosenbrock_function):

$$f(x,y) = (1-x^2) + 100(y-x^2)^2$$

The corresponding optimization module, scipy.optimize, has a routine to accurately evaluate this function (rosen), its derivative (rosen_der), its **Hessian** matrix (rosen_hess), or the product of the latter with a vector (rosen_hess_prod).

Univariate polynomials

Polynomials are defined in SciPy as a NumPy class, poly1d. This class has a handful of methods associated to compute the coefficients of the polynomial (coeffs or simply c), to compute the roots of the polynomial (r), to compute its derivative (deriv), to compute the symbolic integral (integ), and to obtain the degree (order or simply o), as well as a method (variable) that provides a string holding the name of the variable we would like to use in the proper definition of the polynomial (see the example involving P2).

In order to define a polynomial, we must indicate either its coefficients or its roots:

```
>>> import numpy
>>> P1=numpy.poly1d([1,0,1])          # using coefficients
>>> print (P1)
```

The output is as follows:

```
   2
1 x + 1
```

Now let's find roots, order, and derivative of P1:

```
>>> print (P1.r); print (P1.o); print (P1.deriv())
```

The output is as follows:

```
[ 0.+1.j  0.-1.j]
2
2 x
```

Let's use the `poly1d` class:

```
>>> P2=numpy.poly1d([1,1,1], True)      # using roots
>>> print (P2)
```

The output is as follows:

```
   3     2
1 x  - 3 x  + 3 x  - 1
```

Let's use the `poly1d` class with the `variable` method:

```
>>> P2=numpy.poly1d([1,1,1], True, variable='z')
>>> print (P2)
```

The output is as follows:

```
   3     2
1 z  - 3 z  + 3 z  - 1
```

We may evaluate polynomials by treating them either as (vectorized) functions, or with the __call__ method:

```
>>> P1( numpy.arange(10) )              # evaluate at 0,1,...,9
```

The output is as follows:

```
array([ 1,   2,   5, 10, 17, 26, 37, 50, 65, 82])
```

Let's issue the __call__ command:

```
>>> P1.__call__(numpy.arange(10))       # same evaluation
```

The output is as follows:

```
array([ 1,   2,   5, 10, 17, 26, 37, 50, 65, 82])
```

An immediate application of these ideas is to verify the computation of the natural logarithm of *1 + x* used in the preceding example . When *x* is close to zero, the natural logarithm can be approximated by the following formula:

$$ln\left(1+x\right) \approx x - \frac{x^2}{2} \quad if\ x \to 0$$

This expression can be entered and evaluated in Python using the ideas just presented, as follows:

```
>>> import numpy
>>> Px=numpy.poly1d([-(1./2.),1,0])
>>> print(Px)
```

The output is as follows:

```
        2
-0.5 x + 1 x
```

Let's have a look on the value stored in variable a:

```
>>> a=1./10000000000000000.
>>> print(a)
```

The output for value stored in a is as follows:

```
1e-16
```

We now use Px (which contains one-dimensional polynomial form) on a in the following line of code:

```
>>> Px(a)
```

The output is as follows:

```
9.9999999999999998e-17
```

The result is the same as that obtained before using the SciPy function `scipy.special.log1p`, which verifies the computation.

There are also a few routines associated with polynomials: `roots` (to compute zeros), `polyder` (to compute derivatives), `polyint` (to compute integrals), `polyadd` (to add polynomials), `polysub` (to subtract polynomials), `polymul` (to multiply polynomials), `polydiv` (to perform polynomial division), `polyval` (to evaluate polynomials), and `polyfit` (to compute the best fit polynomial of certain order for two given arrays of data).

The usual binary operators +, -, *, and / perform the corresponding operations with polynomials. In addition, once a polynomial is created, any list of values that interacts with them is immediately casted to a polynomial. Therefore, the following four commands are equivalent:

```
>>> P1=numpy.poly1d([1,0,1])
>>> print(P1)
```

The output for the preceding lines of code is as follows:

```
   2
1 x + 1
```

Let's take a look at the following print() command:

```
>>> print(numpy.polyadd(P1, numpy.poly1d([2,1])))
```

The output is as follows:

```
   2
1 x + 2 x + 2
```

Let's take a look at the following print() command:

```
>>> print(numpy.polyadd(P1, [2,1]))
```

The output is as follows:

```
   2
1 x + 2 x + 2
```

Let's take a look at the following print() command:

```
>>> print(P1 + numpy.poly1d([2,1]))
```

The output is as follows:

```
   2
1 x + 2 x + 2
```

Let's take a look at the following print() command:

```
>>> print(P1 + [2,1])
```

The output is as follows:

```
   2
1 x + 2 x + 2
```

Note how the polynomial division offers both the quotient and reminder values, for example:

```
>>> P1/[2,1]
```

The output is as follows:

```
(poly1d([ 0.5 , -0.25]), poly1d([ 1.25]))
```

This can also be written as follows:

$$\frac{x^2+1}{2x+1} = \underbrace{\left(\frac{1}{2}x - \frac{1}{4}\right)}_{quotient} + \frac{\overbrace{5/4}^{reminder}}{2x+1}$$

A family of polynomials is said to be orthogonal with respect to an inner product if for any two polynomials in the family, their inner product is zero. Sequences of these functions are used as the backbone of extremely fast algorithms of quadrature (for numerical integration of general functions). The `scipy.special` module contains the `poly1d` definitions and allows fast evaluation of the families of orthogonal polynomials, such as **Legendre** (`legendre`), **Chebyshev** (`chebyt`, `chebyu`, `chebyc`, and `chebys`), **Jacobi** (`jacobi`), **Laguerre** and its generalized version (`laguerre` and `genlaguerre`), **Hermite** and its normalized version (`hermite` and `hermitenorm`), and **Gegenbauer** (`gegenbauer`). There are also shifted versions of some of them, such as `sh_legendre`, `sh_chebyt`, and so on.

The usual evaluation of polynomials can be improved for orthogonal polynomials, thanks to their rich mathematical structure. In such cases, we never evaluate them with the generic call methods presented previously. Instead, we employ the `eval_` syntax. For example, we use the following command for Jacobi polynomials:

```
>>> eval_jacobi(n, alpha, beta, x)
```

In order to obtain the graph of the Jacobi polynomial of order `n` = `3` for `alpha` = `0` and `beta` = `1`, for a thousand values of x uniformly spaced from -1 to 1, we could issue the following commands:

```
>>> import numpy
>>> import scipy.special
>>> import matplotlib.pyplot as plt
>>> x=numpy.linspace(-1,1,1000)
>>> plt.plot(x,scipy.special.eval_jacobi(3,0,1,x))
>>> plt.show()
```

The output is as follows:

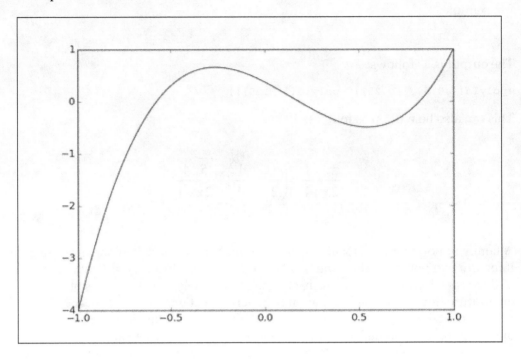

The gamma function

The gamma function is a logarithmic, convex, smooth function operating on complex numbers, which interpolates the factorial function for all nonnegative integers. It is not defined at zero or any negative integer. This is the most common special function and is widely used in many different applications, either by itself or as the main ingredient in the definition of many other functions. The gamma function is used in diverse fields such as quantum physics, astrophysics, statistics, and fluid dynamics.

The gamma function is defined by the improper integral, as follows:

$$\Gamma(z) = \int_0^\infty e^{-t} t^{z-1} \, dt$$

Evaluation of gamma at integer values gives shifted factorials, and that is precisely how the factorials are coded in SciPy.

The `scipy.special` module has algorithms to obtain a fast evaluation of the gamma function at any permissible value. It also contains routines to perform evaluation of the most common compositions of the gamma functions appearing in the literature: `gammaln` for the natural logarithm of the absolute value of gamma, `rgamma` for the value one over gamma, `beta` for quotients, and `betaln` for the natural logarithm of the latter. We also have implementations of the logarithm of its derivative (`psi`).

An obvious application of gamma functions is the ability to perform computations that are virtually impossible for a computer if approached in a direct way. For instance, in statistical applications we often work with ratios of factorials. If these factorials are too large for the precision of a computer, we resort to expressions involving their logarithms instead. Even then, computing *ln(a! / b!)* can prove to be an impossible task (try, for example, with *a = 10**15* and *b = a - 10**10*). An elegant solution uses the digamma function `psi` by an application of the mean value theorem on the `ln(gamma(x))` function. With proper estimation, we obtain the excellent approximation (for this case of choice of *a* and *b*):

$$ln\left(a!/b!\right) \simeq 10^{10}\psi\left(a\right)$$

Let's take a look at the following code snippet:

```
>>> import scipy.special
>>> 10**10*scipy.special.psi(10**15)
```

The output is as follows:

```
345387763949.10681
```

The Riemann zeta function

The Riemann zeta function is very important in analytic number theory and has applications in physics and the probability theory as well. It computes the p-series for any complex value *p*:

$$\zeta\left(p\right) = \sum_{n=1}^{\infty} \frac{1}{n^p}$$

The definition coded in SciPy allows a more flexible generalization of this function, as follows:

$$zeta(a, p) = \sum_{n=0}^{\infty} \frac{1}{(n+a)^p}$$

Among others, this function has applications in the field of particle physics and in dynamical systems (http://en.wikipedia.org/wiki/Hurwitz_zeta_function)

Airy and Bairy functions

These are solutions of the Stokes equation and are obtained by solving the following differential equation:

$$y'' = xy$$

This equation has two linearly independent solutions, both of them defined as an improper integral for real values of the independent variable. The `airy` command computes both functions (`Ai` and `Bi`) as well as their corresponding derivatives (`Aip` and `Bip`, respectively). In the following code, we take advantage of the `contourf` command in `matplotlib.pyplot` to present an image of the real part of the output of the Bairy function `Bi` for an array of 801 x 801 complex values uniformly spaced in the square from *-4 - 4j* to *4 + 4j*. We also offer this graph as a surface plot using the `mplot3d` module of `mpl_toolkits`:

```
>>> import numpy
>>> import scipy.special
>>> import  matplotlib.pyplot as plt
>>> import mpl_toolkits.mplot3d
>>> x=numpy.mgrid[-4:4:100j,-4:4:100j]
>>> z=x[0]+1j*x[1]
>>> (Ai, Aip, Bi, Bip) = scipy.special.airy(z)
>>> steps = range(int(Bi.real.min()), int(Bi.real.max()),6)
>>> fig=plt.figure()
>>> subplot1=fig.add_subplot(121,aspect='equal')
>>> subplot1.contourf(x[0], x[1], Bi.real, steps)
>>> subplot2=fig.add_subplot(122,projection='3d')
>>> subplot2.plot_surface(x[0],x[1],Bi.real)
>>> plt.show()
```

The output is as follows:

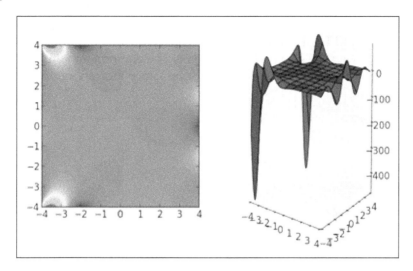

The Bessel and Struve functions

Bessel functions are both of the canonical solutions to Bessel's homogeneous differential equation:

$$x^2 y'' + xy' + \left(x^2 = a^2\right) y = 0$$

These equations arise naturally in the solution of Laplace's equation in cylindrical coordinates. The solutions of the non-homogeneous Bessel differential equation shown in the following diagram are called **Struve** functions:

$$x^2 y'' + xy' + \left(x^2 = a^2\right) y = \frac{4\left(x/2\right)^{a+1}}{\sqrt{\pi}\left(a + \frac{1}{2}\right)}$$

In either case, the order of the equation is the complex number `alpha` which acts as a parameter. Depending on the canonical solution and the order, the Bessel and Struve functions are addressed (and computed) differently.

For Bessel functions, we have algorithms to produce Bessel functions of the first kind (jv) and second kind (yn and yv), Hankel functions of the first and second kind (hankel1 and hankel2), and the modified Bessel functions of the first and second kind (iv, kn, and kv). Their syntax is similar in all cases: first parameter is the order and second parameter the independent variable. The component n in the definition indicates that an integer is to be used as the order (since they are optimally coded for that situation):

```
>>> import numpy
>>> import scipy.special
>>> scipy.special.jn(5,numpy.pi)
```

The output is as follows:

```
0.052141184367118461
```

The scipy.special module also contains fast versions of the most common Bessel functions (those of orders 0 and 1): j0(x), j1(x) (first kind y0(x) and second kind y1(x)), and so on. There are definitions of the spherical Bessel functions, such as sph_jn(n,z) and sph_yn(z); the Riccati-Bessel functions, such as riccati_jn(n,x) and riccati_yn(n,x); and derivatives of all the basic ones, such as jvp, yvp, kvp, ivp, h1vp, and h2vp.

For Struve functions, we have fast algorithms to compute solutions of the differential equation of order v:(struve(v,x) and modstruve(v,x)).

Other special functions

There are more special functions included in the scipy.special module that are of great use in many applications in both pure and applied mathematics. An exhaustive list would be too large for the scope of this chapter, and I encourage you to use the different utilities for each set of special functions. Among the most interesting ones, we have elliptic functions, **Gauss hypergeometric functions**, **parabolic cylinder functions**, **Mathieu functions**, **spheroidal wave functions**, and **Kelvin functions**.

Interpolation

Interpolation is a basic method in numerical computation that is obtained from a discrete set of data points, intended to find an interpolation function which represents some higher order structure that contains the data. The best known example is the interpolation of a sequence of points (x_k and y_k) in a plane to obtain a curve that goes through all the points in the order dictated by the sequence.

If the points in the previous sequence are in the right position and order, it is possible to find a univariate function $y = f(x)$ for which $y\_k = f(x\_k)$. It is often reasonable to request this interpolating function to be a polynomial, or a rational function, or a more complex functional object. Interpolation is also possible in higher dimensions, of course. The objective of the `scipy.interpolate` module is to offer a complete set of optimally coded applications to address this problem in different settings.

Let's address the easiest way of interpolating data to obtain a polynomial: lagrange interpolation. Given a sequence of different x values of size n and a sequence of arbitrary real values y of the same size n, we seek a polynomial $p(x)$ of the degree of $n - 1$ that satisfies the n constraints $p(x[k]) = y[k]$ for all k from 0 to $n - 1$. The following code illustrates how to obtain a polynomial of degree 9 that interpolates the 10 uniformly spaced values of sine in the interval (-1, 1):

```
>>> import numpy
>>> import matplotlib.pyplot as plt
>>> import scipy.interpolate
>>> x=numpy.linspace(-1,1,10); xn=numpy.linspace(-1,1,1000)
>>> y=numpy.sin(x)
>>> polynomial=scipy.interpolate.lagrange(x, numpy.sin(x))
>>> plt.plot(xn,polynomial(xn),x,y,'or')
>>> plt.show()
```

We will obtain the following `plot` showing the Lagrange interpolation:

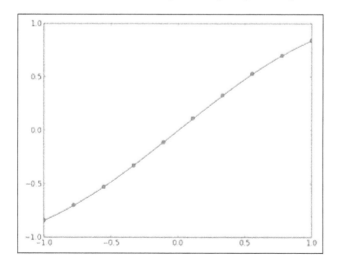

There are numerous issues with Lagrange interpolation. The first obvious drawback is that the user cannot specify the degree of the interpolation; this depends solely on the data. The procedure is also highly unstable numerically, especially for datasets with size over 20 points. This issue can be addressed by allowing the algorithm to depend on different properties of the dataset, rather than just the size and location of the points.

Also, it is inconvenient when we need to update the dataset by adding a few more instances; the procedure needs to be repeated again from the beginning. This proves impractical if the datasets are increasing in size and are updated frequently. To address this issue, `BarycentricInterpolator` has the `add_xi` and `set_yi` methods. For example, in the next session we start by interpolating 10 uniformly spaced values of the sine function between 1 and 10. Once done, we update the interpolating polynomial with 10 more uniformly spaced values between 1.5 and 10.5. As expected, this operation reduces the (percent) relative error of an interpolation computed at points within the interpolating ones. The following commands are used:

```
>>> import numpy
>>> import scipy.interpolate
>>> x1=numpy.linspace(1,10,10); y1=numpy.sin(x1)
>>> Polynomial=scipy.interpolate.BarycentricInterpolator(x1,y1)
>>> exactValues=numpy.sin(x1+0.3)
>>> exactValues
```

Here is the output for `exactValues`:

```
array([ 0.96355819,   0.74570521,  -0.15774569,  -0.91616594,
-0.83226744,
        0.0168139 ,   0.85043662,   0.90217183,   0.12445442,
-0.76768581])
```

Let's find the value of `interpolatedValues` by issuing following commands:

```
>>> interpolatedValues=Polynomial(x1+0.3)
>>> interpolatedValues
```

The output is as follows:

```
array([ 0.97103132,   0.74460631,  -0.15742869,  -0.91631362,
-0.83216445,
        0.01670922,   0.85059283,   0.90181323,   0.12588718,
-0.7825744 ])
```

Let's find the value of `PercentRelativeError` by issuing following commands:

```
>>> PercentRelativeError = numpy.abs((exactValues - interpolatedValues)/
interpolatedValues)*100
>>> PercentRelativeError
```

The output is as follows:

```
array([ 0.76960822,  0.14758101,  0.20136334,  0.01611703,  0.01237594,
        0.62647084,  0.01836479,  0.0397652 ,  1.13812858,  1.90251374])
```

Then, we find what `interpolatedValues2` holds:

```
>>> x2=numpy.linspace(1.5,10.5,10); y2=numpy.sin(x2)
>>> Polynomial.add_xi(x2,y2)
>>> interpolatedValues2=Polynomial(x1+0.3)
>>> interpolatedValues2
```

The output is as follows:

```
array([ 0.96355818,  0.74570521, -0.15774569, -0.91616594, -0.83226744,
        0.0168139 ,  0.85043662,  0.90217183,  0.12445442, -0.76768581])
```

Let's find the value of `PercentRelativeError`, keeping in consideration `interpolatedValues2`:

```
>>> PercentRelativeError = numpy.abs((exactValues - interpolatedValues2)/
interpolatedValues2)*100
>>> PercentRelativeError
```

The output is as follows:

```
array([ 1.26241742e-07,  2.02502252e-09,  5.95225989e-10,
        1.84438143e-11,  8.75086862e-12,  4.14359323e-10,
        1.75194631e-11,  8.52321518e-11,  9.45285176e-09,
        1.29570657e-07])
```

It is possible to interpolate data not only by point location, but also with the derivatives at those locations. The `KroghInterpolator` command allows this by including repeated x values and indicating the location and successive derivatives in order on the corresponding y values.

For instance, if we desire to construct a polynomial that is zero at the origin, one at x = 1, two at x = 2, and has horizontal tangent lines at each of these three locations, we issue the following commands:

```
>>> import numpy
>>> import matplotlib.pyplot as plt
>>> import  scipy.interpolate
>>> x=numpy.array([0,0,1,1,2,2]); y=numpy.array([0,0,1,0,2,0])
>>> interp=scipy.interpolate.KroghInterpolator(x,y)
>>> xn=numpy.linspace(0,2,20)    # evaluate polynomial in larger set
>>> plt.plot(x,y,'o',xn,interp(xn),'r')
>>> plt.show()
```

This renders the following graph:

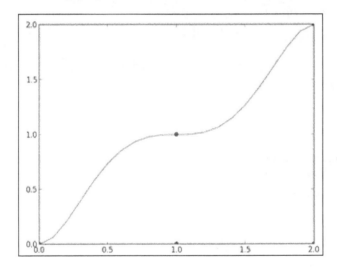

More advanced one-dimensional interpolation is possible with piecewise polynomials (`PiecewisePolynomial`). This allows control over the degrees of different pieces as well as the derivatives at their intersections. Other interpolation options in the `scipy.interpolate` module are **PCHIP monotonic cubic interpolation** (`pchip`) or even **univariate splines** (`InterpolatedUnivariateSpline`).

Let's examine an example with univariate splines. Its syntax is as follows:

```
InterpolatedUnivariateSpline(x, y, w=None, bbox=[None, None], k=3)
```

The x and y arrays contain dependent and independent data, respectively. The array w contains positive weights for spline fitting. The two-sequence bbox parameter specifies the boundary of the approximation interval. The last option indicates the degree of the smoothing polynomials (k).

Suppose we want to interpolate five points as shown in the following example. These points are ordered by strictly increasing x values. We need to perform this interpolation with four cubic polynomials (one for every two consecutive points) in such a way that at least the first derivative of each two consecutive pieces agree on their intersection. We will proceed as follows:

```
>>> import numpy
>>> import matplotlib.pyplot as plt
>>>import scipy.interpolate
>>> x=numpy.arange(5); y=numpy.sin(x)
>>> xn=numpy.linspace(0,4,40)
>>> interp=scipy.interpolate.InterpolatedUnivariateSpline(x,y)
>>> plt.plot(x,y,'.',xn,interp(xn))
>>> plt.show()
```

This offers the following plot showing interpolation with univariate splines:

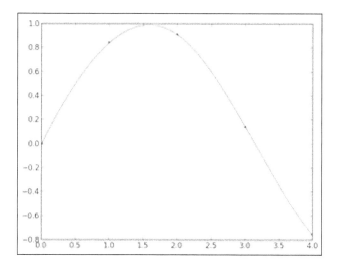

SciPy excels at interpolating in two-dimensional grids as well. It performs well with simple piecewise polynomials (LinearNDInterpolator), piecewise constants (NearestNDInterpolator), or more advanced splines (BivariateSpline). It is capable of carrying out spline interpolation on rectangular meshes in a plane (RectBivariateSpline) or on the surface of a sphere (RectSphereBivariateSpline). For unstructured data, besides the basic scipy.interpolate.BivariateSpline, it is capable of computing smooth approximations (SmoothBivariateSpline) or more involved weighted least-squares splines (LSQBivariateSpline).

The following code creates a 10 x 10 grid of uniformly spaced points in the square from (0, 0) to (9, 9), and evaluates the function sin(x) * cos(y) on the points. We use these points to create a scipy.interpolate.BivariateSpline and evaluate the resulting function on the square for all values:

```
>>> import numpy
>>> import scipy.interpolate
>>> import matplotlib.pyplot as plt
>>> from mpl_toolkits.mplot3d import Axes3D
>>> x=y=numpy.arange(10)
>>> f=(lambda i,j: numpy.sin(i)*numpy.cos(j))   # function to interpolate
>>> A=numpy.fromfunction(f, (10,10))            # generate samples
>>> spline=scipy.interpolate.RectBivariateSpline(x,y,A)
>>> fig=plt.figure()
>>> subplot=fig.add_subplot(111,projection='3d')
>>> xx=numpy.mgrid[0:9:100j, 0:9:100j]       # larger grid for plotting
>>> A=spline(numpy.linspace(0,9,100), numpy.linspace(0,9,100))
>>> subplot.plot_surface(xx[0],xx[1],A)
>>> plt.show()
```

The output is as follows, and it shows the interpolation of 2D data with bivariate splines:

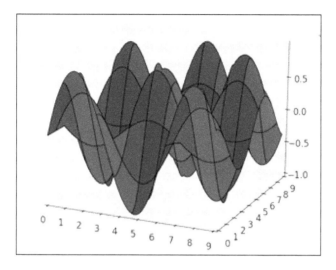

Regression

Regression is similar to interpolation. In this case, we assume that the data is imprecise, and we require an object of predetermined structure to fit the data as closely as possible. The most basic example is univariate polynomial regression to a sequence of points. We obtain that with the `polyfit` command, which we discussed briefly in the *Univariate polynomials* section of this chapter. For instance, if we want to compute the regression line in the least-squares sense for a sequence of 10 uniformly spaced points in the interval $(0, \pi/2)$ and their values under the `sin` function, we will issue the following commands:

```
>>> import numpy
>>> import scipy
>>> import matplotlib.pyplot as plt
>>> x=numpy.linspace(0,1,10)
>>> y=numpy.sin(x*numpy.pi/2)
>>> line=numpy.polyfit(x,y,deg=1)
>>> plt.plot(x,y,'.',x,numpy.polyval(line,x),'r')
>>> plt.show()
```

This gives the following plot that shows linear regression with `polyfit`:

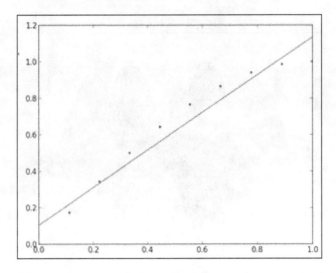

Curve fitting is also possible with splines if we use the parameters wisely. For example, in the case of univariate spline fitting that we introduced before, we can play around with the weights, smoothing factor, the degree of the smoothing spline, and so on. If we want to fit a parabolic spline for the same data as the previous example, we could issue the following commands:

```
>>> import numpy
>>> import scipy.interpolate
>>> import matplotlib.pyplot as plt
>>> x=numpy.linspace(0,1,10)
>>> y=numpy.sin(x*numpy.pi/2)
>>> spline=scipy.interpolate.UnivariateSpline(x,y,k=2)
>>> xn=numpy.linspace(0,1,100)
>>> plt.plot(x,y,'.', xn, spline(xn))
>>> plt.show()
```

This gives the following graph that shows curve fitting with splines:

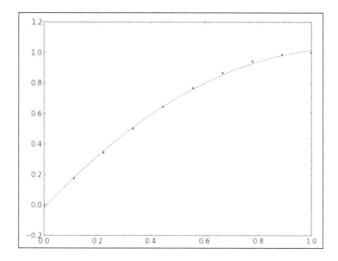

For regression from the point of view of curve fitting, there is a generic routine: `curve_fit` in the `scipy.optimize` module. This routine minimizes the sum of squares of a set of equations using the **Levenberg-Marquardt** algorithm and offers a best fit from any kind of functions (not only polynomials or splines). The syntax is simple:

```
curve_fit(f, xdata, ydata, p0=None, sigma=None, **kw)
```

The `f` parameter is a callable function that represents the function we seek, and `xdata` and `ydata` are arrays of the same length that contain the x and y coordinates of the points to be fit. The tuple `p0` holds an initial guess for the values to be found, and `sigma` is a vector of weights that could be used instead of the standard deviation of the data, if necessary.

We will show its usage with a good example. We will start by generating some points on a section of a sine wave with amplitude A=18, angular frequency $w=3\pi$, and phase h=0.5. We corrupt the data in the array y with some small random noise:

```
>>> import numpy
>>> import scipy
>>> A=18; w=3*numpy.pi; h=0.5
>>> x=numpy.linspace(0,1,100); y=A*numpy.sin(w*x+h)
>>> y += 4*((0.5-scipy.rand(100))*numpy.exp(2*scipy.rand(100)**2))
```

We want to estimate the values of A, w, and h from the corrupted data, hence technically finding a curve fit from the set of sine waves. We start by gathering the three parameters in a list and initializing them to some values, for example, A = 20, $w = 2\pi$, and h = 1. We also construct a callable expression of the target function (target_function):

```
>>> import scipy.optimize
>>> p0 = [20, 2*numpy.pi, 1]
>>> target_function = lambda x,AA,ww,hh: AA*numpy.sin(ww*x+hh)
```

We feed these, together with the fitting data, to curve_fit in order to find the required values:

```
>>> pF,pVar = scipy.optimize.curve_fit(target_function, x, y, p0)
```

A sample of pF run on any of our experiments should give an accurate result for the three requested values:

```
>>> print (pF)
```

The output for the preceding command is as follows:

```
[ 18.13799397    9.32232504    0.54808516]
```

This means that A was estimated to about 18.14, w was estimated very close to 3π, and h was between 0.46 and 0.55. The output of the initial data together with a computation of the sine wave is as follows, in which original data (in blue on the left-hand side graph), corrupted (in red in both graphs), and computed sine wave (in black in the right-hand side) are shown in following plots:

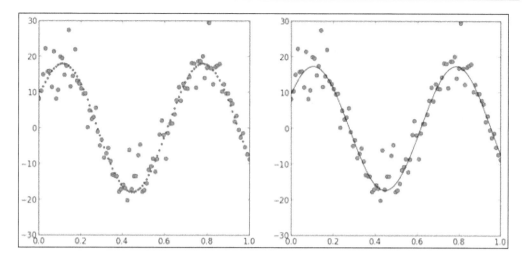

The code is too long to be included here. Instead, the full code (intermediate plots that are produced are not shown here) can be found in the corresponding electronic resource IPython Notebook for this chapter.

Optimization

Optimization involves finding extreme values of functions or their roots. We have already seen the power of optimization in the curve-fitting arena, but it does not stop there. There are applications to virtually every single branch of engineering, and robust algorithms to perform these tasks are a must in every scientist's toolbox.

The `curve_fit` routine is actually syntactic sugar for the general algorithm that performs least-squares minimization, `leastsq`, with the imposing syntax:

```
leastsq(func, x0, args=(), Dfun=None, full_output=0,
        col_deriv=0, ftol=1.49012e-8, xtol=1.49012e-8,
        gtol=0.0, maxfev=0, epsfcn=0.0, factor=100, diag=None):
```

For instance, the `curve_fit` routine could have been called with a `leastsq` call instead:

```
leastsq(error_function,p0,args=(x,y))
```

Here, `error_function` is equal to `lambda p,x,y: target_function(x,p[0],p[1],p[2])-y`

The implementation is given in the corresponding section on the IPython Notebook of this chapter. Most of the optimization routines in SciPy can be accessed from either native Python code, or as wrappers for Fortran or C classical implementations of their corresponding algorithms. Technically, we are still using the same packages we did under Fortran or C, but from within Python. For instance, the minimization routine that implements the truncated `Newton` method can be called with `fmin_ncg` (and this is purely Python) or as `fmin_tnc` (and this one is a wrap of a C implementation).

Minimization

For general minimization problems, SciPy has many different algorithms. So far, we have covered the least-squares algorithm (`leastsq`), but we also have brute force (`brute`), **simulated annealing** (`anneal`), **Brent** or **Golden** methods for scalar functions (`brent` or `golden`), the **downhill simplex** algorithm (`fmin`), **Powell's** method (`fmin_powell`), **nonlinear conjugate gradient** or Newton's version of it (`fmin_cg`, `fmin_ncg`), and the **BFGS** algorithm (`fmin_bfgs`).

Constrained minimization is also possible computationally, and SciPy has routines that implement the **L-BFGS-S** algorithm (`fmin_l_bfgs_s`), truncated Newton's algorithm (`fmin_tnc`), **COBYLA** (`fmin_cobyla`), or sequential least-squares programming (`fmin_slsqp`).

The following code, for example, compares the output of all different methods to finding a local minimum of the Rosenbrock function, `scipy.optimize.rosen`, near the origin using the downhill simplex algorithm:

```
>>> import scipy.optimize
>>> scipy.optimize.fmin(scipy.optimize.rosen,[0,0])
```

The output is as follows:

```
Optimization terminated successfully.
        Current function value: 0.000000
        Iterations: 79
        Function evaluations: 146
array([ 1.00000439,  1.00001064])
```

Since the Version 0.11 of SciPy, all minimization routines can be called from the generic `scipy.optimize.minimize`, with the `method` parameter pointing to one of the strings, such as `Nelder-Mead` (for the downhill simplex), `Powell`, `CG`, `Newton-CG`, `BFGS`, or `anneal`. For constrained minimization, the corresponding strings are one of `L-BFGS-S`, `TNC` (for truncated Newton's), `COBYLA`, or `SLSQP`:

```
minimize( fun, x0, args=(), method='BFGS', jac=None, hess=None,
    hessp=None, bounds=None, constraints=(),tol=None, callback=None,
    options=None)
```

Roots

For most special functions included in the `scipy.special` module, we have accurate algorithms that allow us to their zeros. For instance, for the Bessel function of first kind with integer order, `jn_zeros`, offers as many roots as desired (in ascending order). We may obtain the first three roots of the Bessel J-function of order four by issuing the following command:

```
>>> import scipy.special
>>> print (scipy.special.jn_zeros(4,3))
```

The output is as follows:

```
[  7.58834243   11.06470949   14.37253667]
```

For nonspecial scalar functions, the `scipy.optimize` module allows approximation to the roots through a great deal of different algorithms. For scalar functions, we have the **crude bisection** method (`bisect`), the **classical secant** method of **Newton-Raphson** (`newton`), and more accurate and faster methods such as **Ridders'** algorithm (`ridder`), and two versions of the Brent method (`brentq` and `brenth`).

Finding roots for functions of several variables is very challenging in many ways; the larger the dimension, the more difficult it is. The effectiveness of any of these algorithms depends on the problem, and it is a good idea to invest some time and resources in knowing them all. Since Version 0.11 of SciPy, it is possible to call any of the designed methods with the same routine `scipy.optimize.root`, which has the following syntax:

```
root(fun, x0, args=(), method='hybr', jac=None, tol=None,
    callback=None, options=None)
```

The different methods are obtained upon changing the value of the method parameter to a method's string. We may choose from methods such as 'hybr' for a modified hybrid Powell's method; 'lm' for a modified least-squares method; 'broyden1' or 'broyden2' for Broyden's good and bad methods, respectively; 'diagbroyden' for the diagonal Broyden Jacobian approximation; 'anderson' for Anderson's extended mixing; 'Krylov' for Krylov approximation of the Jacobian; 'linearmixing' for scalar Jacobian approximation; and 'excitingmixing' for a tuned diagonal Jacobian approximation.

For large-scale problems, both the Krylov approximation of the Jacobian or the Anderson extended mixing are usually the best options.

Let's present an illustrative example of the power of these techniques. Consider the following system of differential equations:

$$\begin{cases} x' = x^2 - 2x - y + 0.5 \\ y' = x^2 + 4y^2 - 4 \end{cases}$$

We use the plot routine quiver from the matplotlib.pyplot libraries to visualize a slope field for values of x and y between -0.5 and 2.5, and hence identify the location of the possible critical points in that region:

```
>>> import numpy
>>> import matplotlib.pyplot as plt
>>> f=lambda x: [x[0]**2 - 2*x[0] - x[1] + 0.5, x[0]**2 + 4*x[1]**2 -
    4]
>>> x,y=numpy.mgrid[-0.5:2.5:24j,-0.5:2.5:24j]
>>> U,V=f([x,y])
>>> plt.quiver(x,y,U,V,color='r', \
        linewidths=(0.2,), edgecolors=('k'), \
        headaxislength=5)
>>> plt.show()
```

The output is as follows:

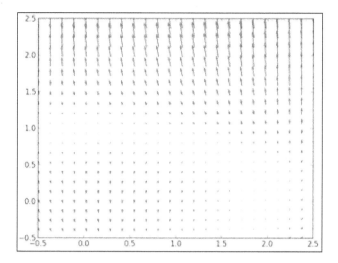

Note how there is a whole region of the plane in which the slopes are extremely small. Because of the degrees of the polynomials involved, there are at most four different possible critical points. In this area, we should be able to identify two such points (as a matter of fact there are only two noncomplex solutions). One of them seems to be near $(0, 1)$ and the second one is near $(2, 0)$. We use these two locations as initial guesses for our searches:

```
>>> import scipy.optimize
>>> f=lambda x: [x[0]**2 - 2*x[0] - x[1] + 0.5, x[0]**2 + 4*x[1]**2 -
    4]
>>> scipy.optimize.root(f,[0,1])
```

The output is as follows:

```
   status: 1
  success: True
qtf: array([ -4.81190247e-09,  -3.83395899e-09])
nfev: 9
      r: array([ 2.38128242, -0.60840482, -8.35489601])
    fun: array([  3.59529073e-12,   3.85025345e-12])
      x: array([-0.22221456,  0.99380842])
```

```
message: 'The solution converged.'
fjac: array([[-0.98918813, -0.14665209],
       [ 0.14665209, -0.98918813]])
```

Let's look at second case:

```
>>> scipy.optimize.root(f,[2,0])
```

The output is as follows:

```
   status: 1
  success: True
qtf: array([  2.08960516e-10,    8.61298294e-11])
nfev: 12
       r: array([-4.56575336, -1.67067665, -1.81464307])
     fun: array([  2.44249065e-15,    1.42996726e-13])
       x: array([ 1.90067673,   0.31121857])
  message: 'The solution converged.'
fjac: array([[-0.39612596, -0.91819618],
       [ 0.91819618, -0.39612596]])
```

In the first case, we converged successfully to (-0.22221456, 0.99380842). In the second case, we converged to (1.90067673, 0.31121857). The routine gives us the details of the convergence and the properties of the approximation. For instance, `nfev` tells us about the number of function calls performed, and `fun` indicates the output of the function at the found location. The other items in the output reflect the matrices used in the procedure, such as `qtf`, `r`, and `fjac`.

Integration

SciPy is capable of performing very robust numerical integration. Definite integrals of a set of special functions are evaluated accurately with routines in the `scipy.special` module. For other functions, there are several different algorithms to obtain reliable approximations in the `scipy.integrate` module.

Exponential/logarithm integrals

A summary of the indefinite and definite integrals in the category of exponential/logarithm is presented here: the exponential integrals (expn, expi, and exp1), **Dawson's** integral (dawsn), and **Gauss error functions** (erf and erfc). We also have **Spence's** dilogarithm (also known as Spence's integral). Let's have a look at the following formulas:

$$\text{expn}\left(n,x\right) = \int_1^\infty \frac{e^{-xt}}{t^n}\,dt \quad \text{exp1}\left(x\right) = \int_1^\infty \frac{e^{-xt}}{t^n}\,dt$$

$$\text{expi}\left(x\right) = \int_{-\infty}^x \frac{e^t}{t}\,dt \quad \text{dawsn}\left(x\right) = e - x^2 \int_0^x e^{t^2}\,dt$$

$$\text{erf}\left(x\right) = \frac{2}{\sqrt{\pi}}\int_0^x e^{-t^2}\,dt \quad \text{erfc}\left(x\right) = \frac{2}{\sqrt{\pi}}\int_x^\infty e^{-t^2}\,dt$$

$$\text{spence}\left(x\right) = -\int_1^x \frac{\log t}{t-1}\,dt$$

Trigonometric and hyperbolic trigonometric integrals

In the category of trigonometric and hyperbolic trigonometric integrals, we have Fresnel sine and cosine integrals, as well as the sinc and hyperbolic trigonometric integrals. Let's have a look at the following formulas:

$$\text{fresnel}\left(z\right) = \int_0^z \sin\left(\frac{\pi}{2}t^2\right)\,dt$$

$$\text{sici}\left(x\right) = \int_0^x \frac{\sin t}{t}\,dt,\ \gamma + \log x + \int_0^x \frac{\cos t - 1}{t}\,dt$$

$$\text{shichi}\left(x\right) = \int_0^x \frac{\sinh t}{t}\,dt,\ \gamma + \log x + \int_0^x \frac{\cosh t - 1}{t}\,dt$$

In the definitions given in the preceding list of integrals, the gamma symbol denotes the Euler-Mascheroni constant:

$$\gamma = \lim_{n \to \infty} \left(\sum_{k=1}^{n} \frac{1}{k} - \log n \right)$$

Elliptic integrals

Elliptic integrals arise naturally when computing the arc length of ellipses. SciPy follows the argument notation for elliptic integrals: complete (one argument) and incomplete (two arguments). Let's have a look at the following formulas:

$$\text{ellipkm1}(m) = \int_{0}^{\pi/2} \frac{d\theta}{\sqrt{1 - m \sin^2 \theta}} \qquad \text{ellipe}(m) = \int_{0}^{\pi/2} \sqrt{1 - m \sin^2 \theta} \, d\theta$$

$$\text{ellipkinc}(m, n) = \int_{0}^{n} \frac{d\theta}{\sqrt{1 - m \sin^2 \theta}} \qquad \text{ellipeinc}(m, n) = \int_{0}^{n} \sqrt{1 - m \sin^2 \theta} \, d\theta$$

Gamma and beta integrals

In the category of gamma and beta integrals, we have one incomplete gamma function, one complemented incomplete gamma integral, and one incomplete beta integral. These are some of the most useful functions in this category. Let's have a look at the following formulas:

$$\text{gammainc}(a, x) = \frac{1}{\Gamma(a)} \int_{0}^{x} e^{-t} t^{a-1} \, dt$$

$$\text{gammaincc}(a, x) = \frac{1}{\Gamma(a)} \int_{x}^{\infty} e^{-t} t^{a-1} \, dt$$

$$\text{betainc}(a, b, c) = \frac{\Gamma(a+b)}{\Gamma(a)\Gamma(b)} \int_{0}^{x} t^{a-1} (t-1)^{b-1} \, dt$$

Numerical integration

For any other functions, we are content with approximating definite integrals with quadrature formulae, such as `quad` (adaptive quadrature), `fixed_quad` (fixed-order Gaussian quadrature), `quadrature` (fixed-tolerance Gaussian quadrature), and `romberg`, (Romberg integration). For functions with more than one variable, we have `dbquad` (double integral) and `tplquad` (triple integral) methods. The syntax in all cases is a variation of `quad`:

```
quad(func, a, b, args=(), full_output=0, epsabs=1.49e-08,
     epsrel=1.49e-08, limit=50, points=None, weight=None,
     wvar=None, wopts=None, maxp1=50, limlst=50)
```

If we have samples instead of functions, we may use routines such as `trapz`, `cumtrapz` (composite trapezoidal rule and its cumulative version), `romb` (Romberg integration again), and `simps` (Simpson's rule) instead. In these routines, the syntax is simpler and changes the order of the parameters. For example, this is how we call `simps`:

```
>>> simps(y, x=None, dx=1, axis=-1, even='avg')
```

Those of us familiar with the **QUADPACK** libraries will find similar syntax, usage, and performance.

For extra information, run the `scipy.integrate.quad_explain()` command. In the IPython Notebook for this chapter, the alternative help command, `scipy.integrate.quad`, is executed and its output is displayed in the corresponding section. This explains with great detail all the different outputs of the quadrature integrals included in the module result, the estimation of absolute error and convergence, and explanation of the used weightings, if necessary. Let's give at least one meaningful example where we integrate a special function and compare the output of a quadrature formula against the more accurate value of the routines given in `scipy.special`:

```
>>> f=lambda t: numpy.exp(-t)*t**4
>>> from scipy.special import gammainc
>>> from scipy.integrate import quad
>>> from scipy.misc import factorial
>>> print (gammainc(5,1))
```

The output is as follows:

```
0.00365984682734
```

Let's take a look at following `print` command:

```
print('%.19f' % gammainc(5,1))
```

The output is as follows:

```
0.0036598468273437131
```

Let's look further into the code:

```
>>> import numpy
>>> result,error=quad(f,0,1)/factorial(4)
>>> result
```

The output is as follows:

```
0.0036598468273437122
```

To use a routine that integrates from samples, we have the flexibility of assigning the frequency and length of the data. For the following problem, we could try with 10,000 samples in the same interval:

```
>>> import numpy
>>> import scipy.integrate
>>> x=numpy.linspace(0,1,10000)
>>> scipy.integrate.simps(f(x)/factorial(4), x)
```

The output is as follows:

```
0.0036598468273469071
```

Ordinary differential equations

As with integration, SciPy has some extremely accurate general-purpose solvers for systems of ordinary differential equations of first order:

$$\frac{dy}{dt} = f(t,y), \quad y(t) = \left(y_1(t),...,y_n(t)\right), t \in \mathbb{R}$$

For real-valued functions, we have basically two flavors: ode (with options passed with the set_integrator method) and odeint (simpler interface). The syntax of ode is as follows:

```
ode(f,jac=None)
```

The first parameter, f, is the function to be integrated, and the second parameter, jac, refers to the matrix of partial derivatives with respect to the dependent variables (the Jacobian). This creates an ode object, with different methods to indicate the algorithm to solve the system (set_integrator), the initial conditions (set_initial_value), and different parameters to be sent to the function or its Jacobian.

The options for integration algorithm are 'vode' for real-valued variable coefficient ODE solver, with fixed-leading-coefficient implementation (it provides Adam's method for non-stiff problems and BDF for stiff); 'zvode' for complex-valued variable coefficient ODE solver, with similar options as the preceding option; 'dopri5' for a **Runge-Kutta** method of order (4)5; 'dop853' for a Runge-Kutta method of order 8(5, 3).

The next code snippet presents an example of usage of the scipy.integrate.ode to solve the initial value problem using the following formula:

$$y' = -20y, \quad y(0) = 1$$

We compute each step sequentially and compare it with the actual solution, which is known. You will notice that virtually there is no difference:

```
>>> import numpy
>>> from scipy.integrate import ode
>>> f=lambda t,y: -20*y          # The ODE
>>> actual_solution=lambda t:numpy.exp(-20*t)  # actual solution
>>> dt=0.01               # time step
>>> solver=ode(f).set_integrator('dop853')  # solver
>>> solver.set_initial_value(1,0)       # initial value
>>> while solver.successful() and solver.t<=1+dt:
    # solve the equation at succesive time steps,
    # until the time is greater than 1
    # but make sure that the solution is successful
```

```
    print (solver.t, solver.y, actual_solution(solver.t))
  # We compare each numerical solution with the actual
  # solution of the ODE
    solver.integrate(solver.t + dt)     # solve next step
```

Once run, the preceding code gives us the following output:

```
<scipy.integrate._ode.ode at 0x10eac5e50>
0 [ 1.] 1.0
0.01 [ 0.81873075] 0.818730753078
0.02 [ 0.67032005] 0.670320046036
0.03 [ 0.54881164] 0.548811636094
0.04 [ 0.44932896] 0.449328964117
0.05 [ 0.36787944] 0.367879441171
0.06 [ 0.30119421] 0.301194211912
0.07 [ 0.24659696] 0.246596963942
0.08 [ 0.20189652] 0.201896517995
0.09 [ 0.16529889] 0.165298888222
0.1 [ 0.13533528] 0.135335283237

     ...

0.9 [   1.52299797e-08] 1.52299797447e-08
0.91 [  1.24692528e-08] 1.24692527858e-08
0.92 [  1.02089607e-08] 1.02089607236e-08
0.93 [  8.35839010e-09] 8.35839010137e-09
0.94 [  6.84327102e-09] 6.84327102222e-09
0.95 [  5.60279644e-09] 5.60279643754e-09
0.96 [  4.58718175e-09] 4.58718174665e-09
0.97 [  3.75566677e-09] 3.75566676594e-09
0.98 [  3.07487988e-09] 3.07487987959e-09
0.99 [  2.51749872e-09] 2.51749871944e-09
1.0  [  2.06115362e-09] 2.06115362244e-09
```

The full output is displayed on the corresponding section of the IPython Notebook for this chapter. For systems of differential equations of first order with complex-valued functions, we have a wrapper of ode, which we call with the complex_ode command. Syntax and usage are similar to those of ode.

The syntax of `odeint` is much more intuitive, and more Python friendly:

```
odeint(func, y0, t, args=(), Dfun=None, col_deriv=0,
full_output=0, ml=None, mu=None, rtol=None, atol=None, tcrit=None,
h0=0.0, hmax=0.0, hmin=0.0, ixpr=0, mxstep=0, mxhnil=0, mxordn=12,
mxords=5, printmessg=0)
```

The most impressive part of this routine is that one is able to indicate not only the Jacobian, but also whether this is banded and how many nonzero diagonals are under or over the main diagonal we have (with the `ml` and `mu` options). This speeds up computations by a huge factor. Another amazing feature of `odeint` is the possibility to indicate critical points for the integration (`tcrit`).

We will now introduce an application to analyze Lorentz attractors with the routines presented in this section.

Lorenz attractors

No book on scientific computing is complete without revisiting Lorenz attractors; SciPy excels both at computation of solutions and presentation of ideas based upon systems of differential equations, of course, and we will show how and why in this section.

Consider a two-dimensional fluid cell that is heated from underneath and cooled from above, much like what occurs with the Earth's atmosphere. This creates convection that can be modeled by a single partial differential equation, for which a decent approximation has the form of the following system of ordinary differential equations:

$$\begin{cases} \dfrac{dx}{dt} = \sigma\left(y - x\right) \\[2mm] \dfrac{dy}{dt} = rx - y - xz \\[2mm] \dfrac{dz}{dt} = xy - bz \end{cases}$$

The variable x represents the rate of convective overturning. The variables y and z stand for the horizontal and vertical temperature variations, respectively. This system depends on four physical parameters, the descriptions of which are far beyond the scope of this book. The important point is that we may model Earth's atmosphere with these equations, and in that case a good choice for the parameters is given by `sigma = 10.0`, and `b = 8/3.0`. For certain values of the third parameter, we have systems for which the solutions behave chaotically. Let's explore this effect with the help of SciPy.

In the following code snippet, we will use one of the solvers in the `scipy.integrate` module as well as the plotting utilities:

```
>>> import numpy
>>> from numpy import linspace
>>> import scipy
>>> from scipy.integrate import odeint
>>> import matplotlib.pyplot as plt
>>> from mpl_toolkits.mplot3d import Axes3D
>>> sigma=10.0
>>> b=8/3.0
>>> r=28.0
>>> f = lambda x,t: [sigma*(x[1]-x[0]), r*x[0]-x[1]-x[0]*x[2], x[0]*x[1]-b*x[2]]
```

Let's choose a time interval `t` large enough with a sufficiently dense partition and any initial condition, `y0`. Then, issue the following commands:

```
>>> t=linspace(0,20,2000); y0=[5.0,5.0,5.0]
>>> solution=odeint(f,y0,t)
>>> X=solution[:,0]; Y=solution[:,1]; Z=solution[:,2]
```

If we want to plot a 3D rendering of the solution obtained, we can do so as follows:

```
>>> import matplotlib.pyplot as plt
>>> plt.gca(projection='3d'); plt.plot(X,Y,Z)
>>> plt.show()
```

This produces the following graph that shows a Lorenz attractor:

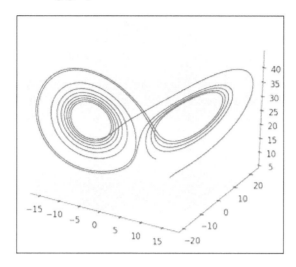

This is most illustrative and shows precisely the chaotic behavior of the solutions. Let's observe the fluctuations of the vertical temperature in detail, along with the fluctuation of horizontal temperature against vertical. Issue the following commands:

```
>>> plt.rcParams['figure.figsize'] = (10.0, 5.0)
>>> plt.subplot(121); plt.plot(t,Z)
>>> plt.subplot(122); plt.plot(Y,Z)
>>> plt.show()
```

This produces the following the plots that show vertical temperature with respect to time (left-hand side plot) and horizontal versus vertical temperature (right-hand side plot):

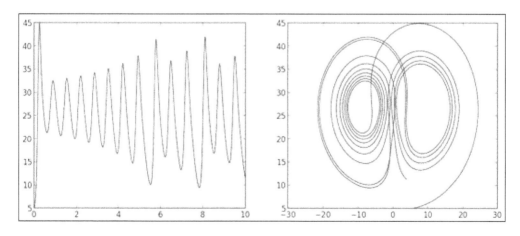

Summary

This chapter explored special functions, integration, interpolation, and optimization through the corresponding modules (`special`, `integrate`, `interpolate`, and `optimize`), as well as discussed solutions of systems of ordinary differential equations.

In *Chapter 5*, *SciPy for Signal Processing*, we will describe the functionality of SciPy modules to analyze processes involving time series and spatial signals, including how to perform on numerical data the discrete Fourier transform, how to construct signals, how to apply filters on data, and how to interpolate images.

5
SciPy for Signal Processing

We define a signal as data that measures either time-varying or spatially varying phenomena. Sound or electrocardiograms are excellent examples of time-varying quantities, while images embody the quintessential spatially varying cases. Moving images (movies or videos) are treated with the techniques of both types of signals, obviously.

The field of signal processing treats four aspects of this kind of data – its acquisition, quality improvement, compression, and feature extraction. SciPy has many routines to treat tasks effectively in any of the four fields. All these are included in two low-level modules (`scipy.signal` being the main one, with an emphasis in time-varying data, and `scipy.ndimage`, for images). Many of the routines in these two modules are based on Discrete Fourier Transform of the data.

In this chapter, we will cover the following things:

- Definition of background algorithms, `scipy.fftpack`
- Built-in functions for signal construction
- Presentation of functions to filter spatial or time series signals

Additional details on the subject can be found in *Python for Signal Processing*, *Unpingco José, Springer Publishing*.

Discrete Fourier Transforms

Discrete Fourier Transform (DFT) transforms any signal from its time/space domain into a related signal in frequency domain. This allows us not only to analyze the different frequencies of the data, but also enables faster filtering operations, when used properly. It is possible to turn a signal in frequency domain back to its time/spatial domain, thanks to the Inverse Fourier Transform (IFT). We will not go into details of the mathematics behind these operators, since we assume familiarity at some level with this theory. We will focus on syntax and applications instead.

The basic routines in the `scipy.fftpack` module compute the DFT and its inverse, for discrete signals in any dimension – `fft`, `ifft` (one dimension); `fft2`, `ifft2` (two dimensions); `fftn`, `ifftn` (any number of dimensions). All of these routines assume that the data is complex valued. If we know beforehand that a particular dataset is actually real valued, and should offer real-valued frequencies, we use `rfft` and `irfft` instead, for a faster algorithm. All these routines are designed so that composition with their inverses always yields the identity. The syntax is the same in all cases, as follows:

```
fft(x[, n, axis, overwrite_x])
```

The first parameter, x, is always the signal in any array-like form. Note that `fft` performs one-dimensional transforms. This means that if x happens to be two-dimensional, for example, `fft` will output another two-dimensional array where each row is the transform of each row of the original. We can use columns instead, with the optional parameter, `axis`. The rest of the parameters are also optional; n indicates the length of the transform and `overwrite_x` gets rid of the original data to save memory and resources. We usually play with the integer n when we need to pad the signal with zeros, or truncate it. For a higher dimension, n is substituted by `shape` (a tuple), and `axis` by `axes` (another tuple).

To better understand the output, it is often useful to shift the zero frequencies to the center of the output arrays with `fftshift`. The inverse of this operation, `ifftshift`, is also included in the module. The following code shows some of these routines in action when applied to a checkerboard image:

```
>>> import numpy
>>> from scipy.fftpack import fft,fft2, fftshift
>>> import matplotlib.pyplot as plt
>>> B=numpy.ones((4,4)); W=numpy.zeros((4,4))
>>> signal = numpy.bmat("B,W;W,B")
>>> onedimfft = fft(signal,n=16)
>>> twodimfft = fft2(signal,shape=(16,16))
>>> plt.figure()
>>> plt.gray()
>>> plt.subplot(121,aspect='equal')
>>> plt.pcolormesh(onedimfft.real)
```

```
>>> plt.colorbar(orientation='horizontal')
>>> plt.subplot(122,aspect='equal')
>>> plt.pcolormesh(fftshift(twodimfft.real))
>>> plt.colorbar(orientation='horizontal')
>>> plt.show()
```

Note how the first four rows of the one-dimensional transform are equal (and so are the last four), while the two-dimensional transform (once shifted) presents a peak at the origin and nice symmetries in the frequency domain.

In the following screenshot, which has been obtained from the previous code, the image on the left is fft and the one on the right is fft2 of a 2 x 2 checkerboard signal:

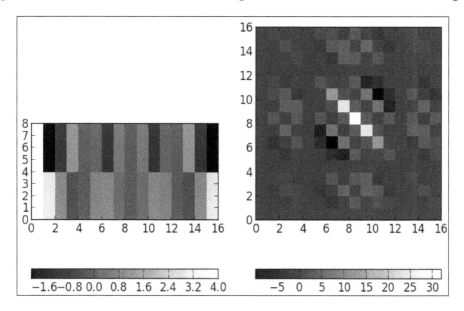

The scipy.fftpack module also offers the Discrete Cosine Transform with its inverse (dct, idct) as well as many differential and pseudo-differential operators defined in terms of all these transforms – diff (for derivative/integral); hilbert, ihilbert (for the Hilbert transform); tilbert, itilbert (for the h-Tilbert transform of periodic sequences); and so on.

Signal construction

To aid the construction of signals with predetermined properties, the
`scipy.signal` module has a nice collection of the most frequent one-dimensional
waveforms in the literature – `chirp` and `sweep_poly` (for the frequency-swept
cosine generator), `gausspulse` (a Gaussian modulated sinusoid), `sawtooth` and
`square` (for the waveforms with those names). They all take as their main parameter
a one-dimensional `ndarray` representing the times at which the signal is to be
evaluated. Other parameters control the design of the signal according to frequency
or time constraints. Let's take a look into the following code snippet which illustrates
the use of these one dimensional waveforms that we just discussed:

```
>>> import numpy
>>> from scipy.signal import chirp, sawtooth, square, gausspulse
>>> import matplotlib.pyplot as plt
>>> t=numpy.linspace(-1,1,1000)
>>> plt.subplot(221); plt.ylim([-2,2])
>>> plt.plot(t,chirp(t,f0=100,t1=0.5,f1=200))    # plot a chirp
>>> plt.title("Chirp signal")
>>> plt.subplot(222); plt.ylim([-2,2])
>>> plt.plot(t,gausspulse(t,fc=10,bw=0.5))       # Gauss pulse
>>> plt.title("Gauss pulse")
>>> plt.subplot(223); plt.ylim([-2,2])
>>> t*=3*numpy.pi
>>> plt.plot(t,sawtooth(t))                      # sawtooth
>>> plt.xlabel("Sawtooth signal")
>>> plt.subplot(224); plt.ylim([-2,2])
>>> plt.plot(t,square(t))                        # Square wave
>>> plt.xlabel("Square signal")
>>> plt.show()
```

Generated by this code, the following diagram shows waveforms for `chirp`, `gausspulse`, `sawtooth`, and `square`:

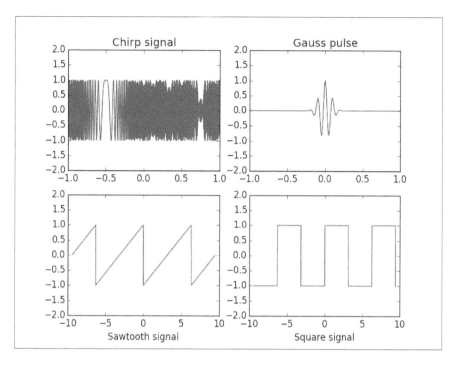

The usual method of creating signals is to import them from a file. This is possible by using purely NumPy routines; for example, `fromfile`:

```
fromfile(file, dtype=float, count=-1, sep='')
```

The `file` argument may point to either a file or a string, the `count` argument is used to determine the number of items to read, and `sep` indicates what constitutes a separator in the original file/string. For images, we have the versatile routine, `imread` in in either the `scipy.ndimage` or `scipy.misc` module:

```
imread(fname, flatten=False)
```

The `fname` argument is a string containing the location of an image. The routine infers the type of file, and reads the data into array accordingly. In case if the `flatten` argument is turned to `True`, the image is converted to gray scale. Note that, in order for `fromfile` and `imread` to work, the **Python Imaging Library** (PIL) needs to be installed.

It is also possible to load `.wav` files for analysis, with the `read` and `write` routines from the `wavfile` submodule in the `scipy.io` module. For instance, the following line of code reads an audio file, say `audio.wav`, using the `read` routine:

```
>>> rate,data = scipy.io.wavfile.read("audio.wav")
```

The command assigns an integer value to the `rate` variable, indicating the sample rate of the file (in samples per second), and a NumPy `ndarray` to the `data` variable, containing the numerical values assigned to the different notes. If we wish to write some one-dimensional `ndarray data` into an audio file of this kind, with the sample rate given by the `rate` variable, we may do so by issuing the following command:

```
>>> scipy.io.wavfile.write("filename.wav",rate,data)
```

Filters

A filter is an operation on signals that either removes features or extracts some component. SciPy has a complete set of known filters as well as the tools to allow construction of new ones. The complete list of filters in SciPy is long, and we encourage the reader to explore the help documents of the `scipy.signal` and `scipy.ndimage` modules for the complete picture. We will introduce in these pages, as an exposition, some of the most used filters in the treatment of audio or image processing.

We start by creating a signal worth filtering:

```
>>> from numpy import sin, cos, pi, linspace
>>> f=lambda t: cos(pi*t) + 0.2*sin(5*pi*t+0.1) + 0.2*sin(30*pi*t) +
0.1*sin(32*pi*t+0.1) + 0.1*sin(47* pi*t+0.8)
>>> t=linspace(0,4,400); signal=f(t)
```

First, we test the classical smoothing filter of **Wiener** and **Kolmogorov**, `wiener`. We present in a `plot` the original signal (in black) and the corresponding filtered data, with a choice of Wiener window of size 55 samples (in blue). Next we compare the result of applying the median filter, `medfilt`, with a kernel of the same size as before (in red):

```
>>> from scipy.signal import wiener, medfilt
>>> import matplotlib.pylab as plt
>>> plt.plot(t,signal,'k', label='The signal')
>>> plt.plot(t,wiener(signal,mysize=55),'r',linewidth=3, label='Wiener
filtered')
```

```
>>> plt.plot(t,medfilt(signal,kernel_size=55),'b',linewidth=3,
label='Medfilt filtered')
>>> plt.legend()
>>> plt.show()
```

This gives us the following graph showing the comparison of smoothing filters (**Wiener**, in red, is the one that has its starting point just above **0.5** and **Medfilt**, in blue, has its starting point just below **0.5**):

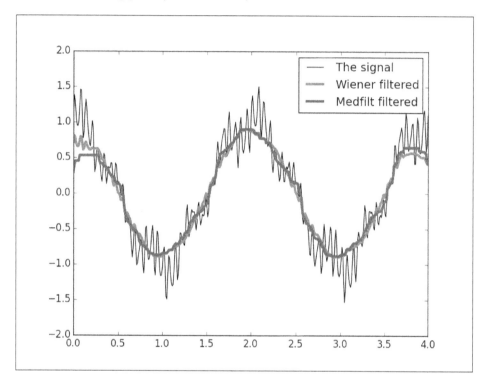

Most of the filters in the `scipy.signal` module can be adapted to work with arrays of any dimension. But in the particular case of images, we prefer to use the implementations in the `scipy.ndimage` module, since they are coded with these objects in mind. For instance, to perform a median filter on an image for smoothing, we use `scipy.ndimage.median_filter`. Let us show an example. We will start by loading Lena to array, and corrupting the image with Gaussian noise (zero mean and standard deviation of 16):

```
>>> from scipy.stats import norm      # Gaussian distribution
>>> import matplotlib.pyplot as plt
```

```
>>> import scipy.misc
>>> import scipy.ndimage
>>> plt.gray()
>>> lena=scipy.misc.lena().astype(float)
>>> plt.subplot(221);
>>> plt.imshow(lena)
>>> lena+=norm(loc=0,scale=16).rvs(lena.shape)
>>> plt.subplot(222);
>>> plt.imshow(lena)
>>> denoised_lena = scipy.ndimage.median_filter(lena,3)
>>> plt.subplot(224);
>>> plt.imshow(denoised_lena)
```

The set of filters for images come in two flavors – statistical and morphological. For example, among the filters of statistical nature, we have the **Sobel** algorithm oriented to detection of edges (singularities along curves). Its syntax is as follows:

```
sobel(image, axis=-1, output=None, mode='reflect', cval=0.0)
```

The optional parameter, `axis`, indicates the dimension in which the computations are performed. By default, this is always the last axis (-1). The `mode` parameter, which is one of the strings `'reflect'`, `'constant'`, `'nearest'`, `'mirror'`, or `'wrap'`, indicates how to handle the border of the image in case there is insufficient data to perform the computations there. In case `mode` is `'constant'`, we may indicate the value to use in the border with the `cval` parameter. Let's look into the following code snippet which illustrates the use of `sobel` filter:

```
>>> from scipy.ndimage.filters import sobel
>>> import numpy
>>> lena=scipy.misc.lena()
>>> sblX=sobel(lena,axis=0); sblY=sobel(lena,axis=1)
>>> sbl=numpy.hypot(sblX,sblY)
>>> plt.subplot(223);
>>> plt.imshow(sbl)
>>> plt.show()
```

The following screenshot illustrates the previous two filters in action—Lena (upper-left), noisy Lena (upper-right), edge map with sobel (lower-left), and median filter (lower-right):

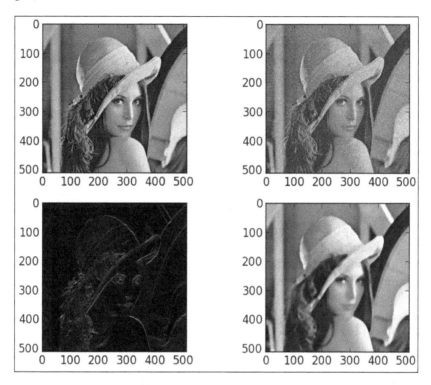

The LTI system theory

To investigate the response of a time-invariant linear system to input signals, we have many resources in the `scipy.signal` module. As a matter of fact, to simplify representation of objects, we have an `lti` class (linear-time invariant class) with associated methods such as `bode` (to calculate bode magnitude and phase data), `impulse`, `output`, and `step`.

Whether we are working with continuous or discrete-time linear systems, we have routines to simulate such systems (`lsim` and `lsim2` for continuous, `dsim` for discrete), as well as compute impulses (`impulse` and `impulse2` for continuous, `dimpulse` for discrete) and steps (`step` and `step2` for continuous, `dstep` for discrete).

Transforming a system from continuous to discrete is possible with `cont2discrete`, but in either case we are able to provide for any system with any of its representations, as well as to convert from one to another. For instance, if we have the zeros `z`, poles `p`, and system gain `k` of the transfer function, we may obtain the polynomial representation (numerator first, then denominator) with `zpk2tf(z,p,k)`. If we have numerator (`num`) and denominator (`dem`) of the transfer function, we obtain the state-space with `tf2ss(num,dem)`. This operation is reversible with the `ss2tf` routine. The change of representation from zero-pole-gain to/from state-space is also contemplated in the (`zpk2ss`, `ss2zpk`) module.

Filter design

There are routines in the `scipy.signal` module that allow the creation of different kinds of filters with diverse methods. For instance, the `bilinear` function returns a digital filter from an analog using a bilinear transform. **Finite impulse response (FIR)** filters can be designed by the window method with the `firwin` and `firwin2` routines. **Infinite impulse response (IIR)** filters can be designed in two different ways, via `iirdesign` or `iirfilter`. **Butterworth** filters can be designed with the `butter` routine. There are also routines to design filters of **Chebyshev** (`cheby1`, `cheby2`), **Cauer** (`ellip`), and Bessel (`bessel`).

Window functions

No signal processing computational system would be complete without an extensive list of windows — mathematical functions that are zero valued outside specific domains. In this section, we will use a few of the coded windows implemented in the `scipy.signal` module to design very simple smoothing filters by using convolution.

We will be testing them on the same one-dimensional signal we employed before, for comparison.

We start by showing the plot of four well-known window functions – Boxcar, Hamming, Blackman-Harris (Nuttall version), and triangular. We will use a size of 31 samples:

```
>>> from scipy.signal import boxcar, hamming, nuttall, triang
>>> import matplotlib.pylab as plt
>>> windows=['boxcar', 'hamming', 'nuttall', 'triang']
>>> plt.subplot(121)
>>> for w in windows:
```

```
eval( 'plt.plot(' + w + '(31))' )
plt.ylim([-0.5,2]); plt.xlim([-1,32])
plt.legend(windows)
```

We need to extend the original signal by fifteen samples for plotting purposes:

```
>>> plt.subplot(122)
>>> import numpy
>>> from numpy import sin, cos, pi, linspace
>>> f=lambda t: cos(pi*t) + 0.2*sin(5*pi*t+0.1) + 0.2*sin(30*pi*t) +
0.1*sin(32*pi*t+0.1) + 0.1*sin(47* pi*t+0.8)
>>> t=linspace(0,4,400); signal=f(t)
>>> extended_signal=numpy.r_[signal[15:0:-1],signal,signal[-1:-15:- 1]]
>>> plt.plot(extended_signal,'k')
```

The final step is the filter itself, which we perform by a simple convolution:

```
>>> for w in windows:
        window = eval( w+'(31)')
        output=numpy.convolve(window/window.sum(),signal)
        plt.plot(output,linewidth=2)
        plt.ylim([-2,3]); plt.legend(['original']+windows)
>>> plt.show()
```

This produces the following output, showing convolution of a signal with different windows:

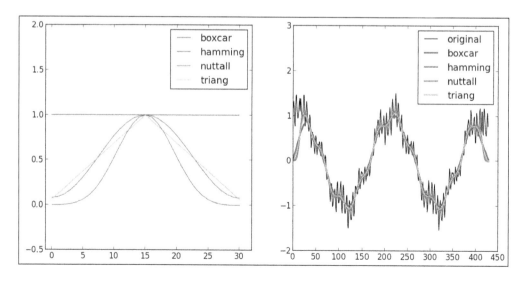

Image interpolation

The set of filters on images that performs some geometric manipulation of the input is classically termed image interpolation, since this numerical technique is the root of all the algorithms. As a matter of fact, SciPy collects all these under the submodule, `scipy.ndimage.interpolation`, for ease of access. This section is best explained through examples, going over the most meaningful routines for geometric transformation. The starting point is the image, Lena. We now assume that all functions from the submodule have been imported into the session.

We need to apply an affine transformation on the domain of the image, given in matrix form as follows:

$$L(x, y) = \underbrace{\begin{pmatrix} a_{11} & a12 \\ a_{21} & a_{22} \end{pmatrix}}_{A} \begin{pmatrix} x \\ y \end{pmatrix} + \underbrace{\begin{pmatrix} b_1 \\ b_2 \end{pmatrix}}_{b}$$

To apply the transformation on the domain of the image we issue the `affine_transform` command (note that the syntax is self-explanatory):

```
>>> import scipy.misc
>>> import numpy
>>> import matplotlib.pylab as plt
>>> from scipy.ndimage.interpolation import affine_transform
>>> lena=scipy.misc.lena()
>>> A=numpy.mat("0,1;-1,1.25"); b=[-400,0]
>>> Ab_Lena=affine_transform(lena,A,b,output_shape=(512*2.2,512*2.2))
>>> plt.gray()
>>> plt.subplot(121)
>>> plt.imshow(Ab_Lena)
```

For a general transformation, we use the `geometric_transform` routine with the following syntax:

```
geometric_transform(input, mapping, output_shape=None,
                    output=None, order=3, mode='constant',
cval=0.0, prefilter=True, extra_arguments=(),
extra_keywords={})
```

We need to provide a rank-2 map from tuples to tuples as the parameter mapping. For instance, we desired to apply the **Möbius** transform for complex-valued number z (where we assume the values of a, b, c, and d are already defined and they are complex-valued numbers) in the following formula:

$$f(z) = \frac{az + b}{cz + d}$$

We would have to code it in the following way:

```
>>> def f(z):
        temp = a*(z[0]+1j*z[1]) + b
        temp /= c*(z[0]+1j*z[1])+d
        return (temp.real, temp.imag)
```

In both functions, the values of the grid that cannot be computed directly with the formula are inferred with spline interpolation. We may specify the order of this interpolation with the `order` parameter. The points outside the domain of definition are not interpolated, but filled according to some predetermined rule. We may impose this rule by passing a string to the `mode` option. The choices are – `'constant'`, to use a constant value that we may impose with the `cval` option; `'nearest'`, that continues the last value of the interpolation on each level line; and `'reflect'` or `'wrap'`, which are self-explanatory.

For example, for the values a = `2**15*(1+1j)`, b = `0`, c = `-2**8*(1-1j*2)`, and d = `2**18-1j*2**14`, we obtain (after imposing the `reflect` mode) the result, as shown just after this line of code:

```
>>> from scipy.ndimage.interpolation import geometric_transform
>>> a = 2**15*(1+1j); b = 0; c = -2**8*(1-1j*2); d = 2**18-1j*2**14
>>> Moebius_Lena = geometric_transform(lena,f,mode='reflect')
>>> plt.subplot(122);
>>> plt.imshow(Moebius_Lena)
>>> plt.show()
```

The following screenshot shows affine transformation (left) and geometric transformation (right):

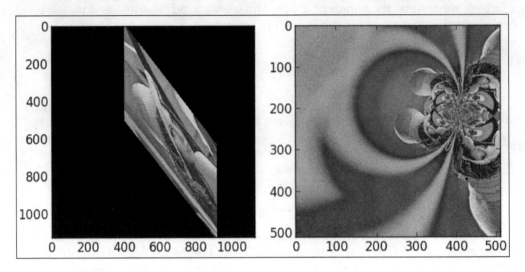

For special cases of rotations, shifts, or dilations, we have the syntactic sugar routines, `rotate(input,angle)`, `shift(input, offset)`, and `zoom(input,dilation_factor)`.

Given any image, we know the value of the array at pixel values (with integer coordinates) in the domain. But what would the corresponding value of a location be without integer coordinates? We may obtain that information with the valuable routine, map_coordinates. Note that the syntax may be confusing, especially with the `coordinates` parameter:

```
map_coordinates(input, coordinates, output=None, order=3,
                mode='constant', cval=0.0, prefilter=True)
```

For instance, if we wish to evaluate Lena at the locations (10.5, 11.7) and (12.3, 1.4), we collect the coordinates as a sequence of sequences; the first internal sequence contains the x values, and the second, the y values. We may specify the order of splines used with `order`, and the interpolation scheme outside of the domain, if needed, as in the previous examples. Let's evaluate Lena at the locations (which we just discussed in our example) using following code snippet:

```
>>> import scipy.misc
>>> from scipy.ndimage.interpolation import map_coordinates
>>> lena=scipy.misc.lena().astype(float)
>>> coordinates=[[10.5, 12.3], [11.7, 1.4]]
>>> map_coordinates(lena, coordinates, order=1)
```

The output is shown as:

```
array([ 157.2 ,  157.42])
```

Further, we evaluate Lena with `order=2` as shown in following line of code:

```
>>> map_coordinates(lena, coordinates, order=2)
```

The output is shown as:

```
array([ 157.80641507,  157.6741489 ])
```

Morphology

We also have the possibility of creating and applying filters to images based on mathematical morphology, both to binary and gray-scale images. The four basic morphological operations are opening (`binary_opening`), closing (`binary_closing`), dilation (`binary_dilation`), and erosion (`binary_erosion`). Note that the syntax of each of these filters is very simple, since we only need two ingredients – the signal to filter and the structuring element to perform the morphological operation. Let's take a look into the general syntax for these morphological operations:

```
binary_operation(signal, structuring_element)
```

We have illustrated the use some of these operations towards an application to obtain the structural model of an oxide, but we will postpone this example until we cover the notions of triangulations and Voronoi diagrams in *Chapter 7, SciPy for Computational Geometry*.

We may use combinations of these four basic morphological operations to create more complex filters for the removal of holes, hit-or-miss transforms (to find the location of specific patterns in binary images), denoising, edge detection, and many more. The module even provides us with some of the most common filters constructed this way. For instance, for the location of the letter e in a text (which we covered in *Chapter 2, Working with the NumPy Array As a First Step to SciPy*, as an application of correlation), we could use the following command instead:

```
>>> binary_hit_or_miss(text, letterE)
```

For comparative purposes, let's apply this command to the example from *Chapter 2, Working with the NumPy Array As a First Step to SciPy*:

```
>>> import numpy
>>> import scipy.ndimage
>>> import matplotlib.pylab as plt
```

```
>>> from scipy.ndimage.morphology import binary_hit_or_miss
>>> text = scipy.ndimage.imread('CHAP_05_input_textImage.png')
>>> letterE = text[37:53,275:291]
>>> HitorMiss = binary_hit_or_miss(text, structure1=letterE, origin1=1)
>>> eLocation = numpy.where(HitorMiss==True)
>>> x=eLocation[1]; y=eLocation[0]
>>> plt.imshow(text, cmap=plt.cm.gray, interpolation='nearest')
>>> plt.autoscale(False)
>>> plt.plot(x,y,'wo',markersize=10)
>>> plt.axis('off')
>>> plt.show()
```

This generates the following output, which the reader should compare with the corresponding one on *Chapter 2, Working with the NumPy Array As a First Step to SciPy*:

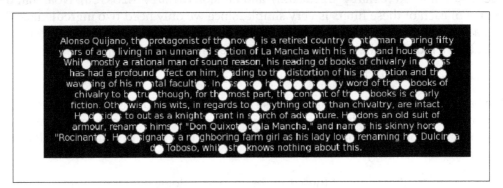

For gray-scale images, we may use a structuring element (`structuring_element`) or a footprint. The syntax is, therefore, a little different:

```
grey_operation(signal, [structuring_element, footprint, size, ...])
```

If we desire to use a completely flat and rectangular structuring element (all *ones*), then it is enough to indicate the size as a tuple. For instance, to perform gray-scale dilation of a flat element of `size` `(15,15)` on our classical image of Lena, we issue the following command:

```
>>> grey_dilation(lena, size=(15,15))
```

The last kind of morphological operations coded in the scipy.ndimage module perform distance and feature transforms. Distance transforms create a map that assigns to each pixel the distance to the nearest object. Feature transforms provide the index of the closest background element instead. These operations are used to decompose images into different labels. We may even choose different metrics such as Euclidean distance, chessboard distance, and **taxicab** distance. The syntax for the distance transform (distance_transform) using a brute force algorithm is as follows:

```
distance_transform_bf(signal, metric='euclidean', sampling=None,
    return_distances=True, return_indices=False,
                    distances=None, indices=None)
```

We indicate the metric with the strings such as 'euclidean', 'taxicab', or 'chessboard'. If we desire to provide the feature transform instead, we switch return_distances to False and return_indices to True.

Similar routines are available with more sophisticated algorithms – distance_transform_cdt (using chamfering for taxicab and chessboard distances). For Euclidean distance, we also have distance_transform_edt. All these use the same syntax.

Summary

In this chapter, we explored signal processing (any dimensional), including the treatment of signals in frequency space, by means of their Discrete Fourier Transforms. These correspond to the fftpack, signal, and ndimage modules.

The *Chapter 6, SciPy for Data Mining*, will explore the tools included in SciPy to approach Statistical and Data Mining problems. In addition to standard statistical quantities, special topics like kernel estimation, statistical distances, and the clustering of big data sets will be presented.

6
SciPy for Data Mining

This chapter covers those branches of mathematics and statistics that treat the collection, organization, analysis, and interpretation of data. There are different applications and operations that spread over several modules and submodules: `scipy.stats` (for purely statistical tools), `scipy.ndimage.measurements` (for analysis and organization of data), `scipy.spatial` (for spatial algorithms and data structures), and finally the clustering package `scipy.cluster`. The `scipy.cluster` clustering package consists of two submodules: `scipy.cluster.vq` (vector quantization) and `scipy.cluster.hierarchy` (for hierarchical and **agglomerative** clustering).

As in the previous chapters, fluency with the subject matter is assumed. Our emphasis is to show you some of the SciPy functions available to perform statistical computations, not to teach it. Accordingly, you are welcome to read this chapter along side your preferred book(s) on the subject so that you can fully explore the examples provided in this chapter on additional data sets.

We should mention, however, that there are other specialized modules in Python that can be used to explore this subject from different perspectives. Some of them (not covered by any means in this book) are the **Modular Toolkit for Data Processing** (**MDP**) (`http://mdp-toolkit.sourceforge.net/install.html`), **scikit-learn** (`http://scikit-learn.org/`), and **Statsmodels** (`http://statsmodels.sourceforge.net/`).

In this chapter, we will cover the following things:

- The standard descriptive statistics measures computed via SciPy
- The built-in functions in SciPy that deal with statistical distributions
- The Scipy functionality to find interval estimation
- Performing computations of statistical correlations and some statistical tests, the fitting of distributions, and statistical distances
- A clustering example

Descriptive statistics

We often require the analysis of data in which certain features are grouped in different regions, each with different sizes, values, shapes, and so on. The `scipy.ndimage.measurements` submodule has the right tools for this task, and the best way to illustrate the capabilities of the module is by means of exhaustive examples. For example, for binary images of zeros and ones, it is possible to label each blob (areas of contiguous pixels with value one) and obtain the number of these with the `label` command. If we desire to obtain the center of mass of the blobs, we may do so with the `center_of_mass command`. We may see these operations in action once again in the application to obtain the structural model of oxides in *Chapter 7, SciPy for Computational Geometry*.

For nonbinary data, the `scipy.ndimage.measurements` submodule provides the usual basic statistical measurements (value and location of extreme values, mean, standard deviation, sum, variance, histogram, and so on).

For more advanced statistical measurements, we must access functions from the `scipy.stats` module. We may now use geometric and harmonic means (`gmean`, `hmean`), median, mode, skewness, various moments, or kurtosis (`median`, `mode`, `skew`, `moment`, `kurtosis`). For an overview of the most significant statistical properties of the dataset, we prefer to use the `describe` routine. We may also compute item frequencies (`itemfreq`), percentiles (`scoreatpercentile`, `percentileofscore`), histograms (`histogram`, `histogram2`), cumulative and relative frequencies (`cumfreq`, `relfreq`), standard error (`sem`), and the signal-to-noise ratio (`signaltonoise`), which is always useful.

Distributions

One of the main strengths of the `scipy.stats` module is the great number of distributions coded, both continuous and discrete. The list is impressively large and has at least 80 continuous distributions and 10 discrete distributions.

One of the most common ways to employ these distributions is the generation of random numbers. We have been employing this technique to *contaminate* our images with noise, for example:

```
>>> import scipy.misc
>>> from scipy.stats import signaltonoise
>>> from scipy.stats import norm       # Gaussian distribution
>>> lena=scipy.misc.lena().astype(float)
>>> lena+= norm.rvs(loc=0,scale=16,size=lena.shape)
>>> signaltonoise(lena,axis=None)
```

The output is shown as follows:

```
array(2.459233897516763)
```

Let's see the SciPy way of handling distributions. First, a random variable class is created (in SciPy there is the `rv_continuous` class for continuous random variables and the `rv_discrete` class for the discrete case). Each continuous random variable has an associated probability density function (`pdf`), a cumulative distribution function (`cdf`), a survival function along with its inverse (`sf`, `isf`), and all possible descriptive statistics. They also have associated the random variable, `rvs`, which is what we used to actually generate the random instances. For example, with a Pareto continuous random variable with parameter $b = 5$, to check these properties, we could issue the following commands:

```
>>> import numpy
>>> from scipy.stats import pareto
>>> import matplotlib.pyplot as plt
>>> x=numpy.linspace(1,10,1000)
>>> plt.subplot(131); plt.plot(pareto.pdf(x,5))
>>> plt.subplot(132); plt.plot(pareto.cdf(x,5))
>>> plt.subplot(133); plt.plot(pareto.rvs(5,size=1000))
>>> plt.show()
```

This gives the following graphs, showing probability density function (left), cumulative distribution function (center), and random generation (right):

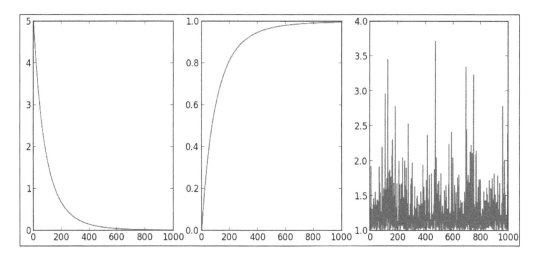

Interval estimation, correlation measures, and statistical tests

We briefly covered interval estimation as an introductory example of SciPy: `bayes_mvs`, in *Chapter 1, Introduction to SciPy*, with very simple syntax, as follows:

```
bayes_mvs(data, alpha=0.9)
```

It returns a tuple of three arguments in which each argument has the form `(center, (lower, upper))`. The first argument refers to the mean; the second refers to the variance; and the third to the standard deviation. All intervals are computed according to the probability given by `alpha`, which is `0.9` by default.

We may use the `linregress` routine to compute the regression line of some two-dimensional data *x*, or two sets of one-dimensional data, *x* and *y*. We may compute different correlation coefficients, with their corresponding p-values, as well. We have the **Pearson correlation coefficient** (`pearsonr`), **Spearman's rank-order correlation** (`spearmanr`), **point biserial correlation** (`pointbiserialr`), and **Kendall's tau** for ordinal data (`kendalltau`). In all cases, the syntax is the same, as it is only required either a two-dimensional array of data, or two one-dimensional arrays of data with the same length.

SciPy also has most of the best-known statistical tests and procedures: **t-tests** (`ttest_1samp` for one group of scores, `ttest_ind` for two independent samples of scores, or `ttest_rel` for two related samples of scores), **Kolmogorov-Smirnov tests** for goodness of fit (`kstest`, `ks_2samp`), one-way **Chi-square test** (`chisquare`), and many more.

Let us illustrate some of the routines of this module with a textbook example, based on Timothy Sturm's studies on control design.

To turn a knob that moved an indicator by the screw action, 25 right-handed individuals were asked to use their right hands. There were two identical instruments, one with a right-handed thread where the knob turned clockwise, and the other with a left-hand thread where the knob turned counter-clockwise. The following table gives the times in seconds each subject took to move the indicator to a fixed distance:

Subject	1	2	3	4	5	6	7	8	9	10
Right thread	113	105	130	101	138	118	87	116	75	96
Left thread	137	105	133	108	115	170	103	145	78	107
Subject	11	12	13	14	15	16	17	18	19	20
Right thread	122	103	116	107	118	103	111	104	111	89
Left thread	84	148	147	87	166	146	123	135	112	93
Subject	21	22	23	24	25					
Right thread	78	100	89	85	88					
Left thread	76	116	78	101	123					

We may perform an analysis that leads to a conclusion about right-handed people finding right-hand threads easier to use, by a simple one-sample t-statistic. We will load the data in memory, as follows:

```
>>> import numpy
>>> data = numpy.array([[113,105,130,101,138,118,87,116,75,96, \
        122,103,116,107,118,103,111,104,111,89,78,100,89,85,88], \
        [137,105,133,108,115,170,103,145,78,107, \
        84,148,147,87,166,146,123,135,112,93,76,116,78,101,123]])
```

The difference of each row indicates which knob was faster, and for how much time. We can obtain that information easily and perform some basic statistical analysis on it. We will start by computing the mean, standard deviation, and a histogram with 10 bins:

```
>>> dataDiff = data[1,:]-data[0,:]
>>> dataDiff.mean(), dataDiff.std()
```

The output is shown as:

```
(13.32, 22.472596645692729)
```

Let's plot the histogram by issuing the following set of commands:

```
>>> import matplotlib.pyplot as plt
>>> plt.hist(dataDiff)
>>> plt.show()
```

This produces the following histogram:

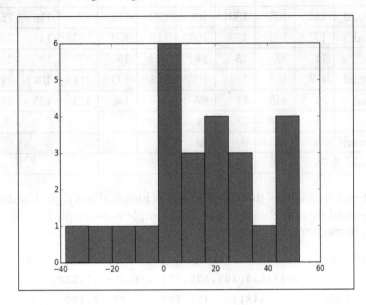

In light of this histogram, it is not far-fetched to assume a normal distribution. If we assume that this is a proper simple random sample, the use of t-statistics is justified. We would like to prove that it takes longer to turn the left thread than the right, so we set the mean of `dataDiff` to be contrasted against the zero mean (which would indicate that it takes the same time for both threads).

The two-sample t-statistics and p-value for the two-sided test are computed by the simple command, as follows:

```
>>> from scipy.stats import ttest_1samp
>>> t_stat,p_value=ttest_1samp(dataDiff,0.0)
```

The p-value for the one-sided test is then calculated:

```
>>> print (p_value/2.0)
```

The output is shown as follows:

```
0.00389575522747
```

Note that this p-value is much smaller than either of the usual thresholds `alpha` = `0.05` or `alpha` = `0.1`. We can thus guarantee that we have enough evidence to support the claim that right-handed threads take less time to turn than left-handed threads.

Distribution fitting

In Timothy Sturm's example, we claim that the histogram of some data seemed to fit a normal distribution. SciPy has a few routines to help us approximate the best distribution to a random variable, together with the parameters that best approximate this fit. For example, for the data in that problem, the mean and standard deviation of the normal distribution that realizes the best fit can be found in the following way:

```
>>> from scipy.stats import norm      # Gaussian distribution
>>> mean,std=norm.fit(dataDiff)
```

We can now plot the (normed) histogram of the data, together with the computed probability density function, as follows:

```
>>> plt.hist(dataDiff, normed=1)
>>> x=numpy.linspace(dataDiff.min(),dataDiff.max(),1000)
>>> pdf=norm.pdf(x,mean,std)
>>> plt.plot(x,pdf)
>>> plt.show()
```

We will obtain the following graph showing the maximum likelihood estimate to the normal distribution that best fits dataDiff:

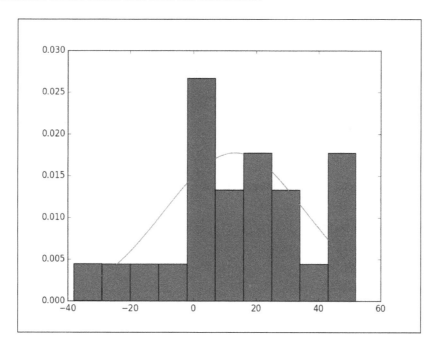

We may even fit the best probability density function without specifying any particular distribution, thanks to a non-parametric technique, **kernel density estimation**. We can find an algorithm to perform Gaussian kernel density estimation in the `scipy.stats.kde` submodule. Let us show by example with the same data as before:

```
>>> from scipy.stats import gaussian_kde
>>> pdf=gaussian_kde(dataDiff)
```

A slightly different plotting session as given before, offers us the following graph, showing probability density function obtained by kernel density estimation on `dataDiff`:

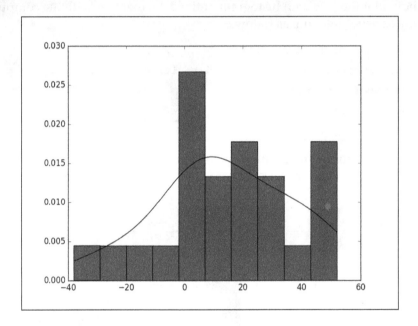

The full piece of code is as follows:

```
>>> from scipy.stats import gaussian_kde
>>> pdf = gaussian_kde(dataDiff)
>>> pdf = pdf.evaluate(x)
>>> plt.hist(dataDiff, normed=1)
>>> plt.plot(x,pdf,'k')
>>> plt.savefig("hist2.png")
>>> plt.show()
```

For comparative purposes, the last two plots can be combined into one:

```
>>> plt.hist(dataDiff, normed=1)
>>> plt.plot(x,pdf,'k.-',label='Kernel fit')
>>> plt.plot(x,norm.pdf(x,mean,std),'r',label='Normal fit')
>>> plt.legend()
>>> plt.savefig("hist3.png")
>>> plt.show()
```

The output is the combined plot as follows:

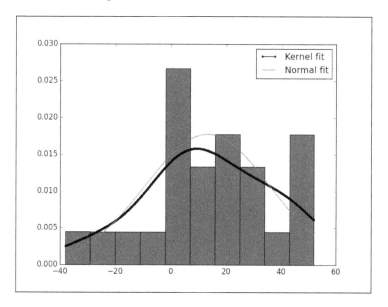

Distances

In the field of data mining, it is often required to determine which members of a training set are closest to unknown test instances. It is imperative to have a good set of different distance functions for any of the algorithms that perform the search, and SciPy has, for this purpose, a huge collection of optimally coded functions in the distance submodule of the scipy.spatial module. The list is long. Besides Euclidean, squared Euclidean, or standardized Euclidean, we have many more—**Bray-Curtis, Canberra, Chebyshev, Manhattan**, correlation distance, cosine distance, **dice dissimilarity, Hamming, Jaccard-Needham, Kulsinski, Mahalanobis**, and so on. The syntax in most cases is simple:

```
distance_function(first_vector, second_vector)
```

The only three cases in which the syntax is different are the Minkowski, Mahalanobis, and standardized Euclidean distances, in which the distance function requires either an integer number (for the order of the norm in the definition of Minkowski distance), a covariance for the Mahalanobis case (but this is an optional requirement), or a variance matrix to standardize the Euclidean distance.

Let us see now a fun exercise to visualize the unit balls in Minkowski metrics:

```
>>> import numpy
>>> from scipy.spatial.distance import minkowski
>>> Square=numpy.mgrid[-1.1:1.1:512j,-1.1:1.1:512j]
>>> X=Square[0]; Y=Square[1]
>>> f=lambda x,y,p: minkowski([x,y],[0.0,0.0],p)<=1.0
>>> Ball=lambda p:numpy.vectorize(f)(X,Y,p)
```

We have created a function, `Ball`, which creates a grid of 512 x 512 Boolean values. The grid represents a square of length 2.2 centered at the origin, with sides parallel to the coordinate axis, and the true values on it represent all those points of the grid inside of the unit ball for the Minkowksi metric, for the parameter p. All we have to do is show it graphically, as in the following example:

```
>>> import matplotlib.pylab as plt
>>> plt.imshow(Ball(3), cmap = plt.cm.gray)
>>> plt.axis('off')
>>> plt.subplots_adjust(left=0.0127,bottom=0.0164,\
    right=0.987,top=0.984)
>>> plt.show()
```

This produces the following, where `Ball(3)` is a unit ball in the Minkowski metric with parameter p = 3:

We feel the need to issue the following four important warnings:

- **First warning**: We must use these routines instead of creating our own definitions of the corresponding distance functions whenever possible. They guarantee a faster result and optimal coding to take care of situations in which the inputs are either too large or too small.

- **Second warning**: These functions work great when comparing two vectors; however, for the pairwise computation of many vectors, we must resort to the `pdist` routine. This command takes an $m \times n$ array representing m vectors of dimension n, and computes the distance of each of them to each other. We indicate the distance function to be used with the option metric and additional parameters as needed. For example, for the Manhattan (`cityblock`) distance for five randomly selected randomly selected four-dimensional vectors with integer values 1, 0, or -1, we could issue the following command:

```
>>> import scipy.stats
>>> from scipy.spatial.distance import pdist
>>> V=scipy.stats.randint.rvs(0.4,3,size=(5,4))-1
>>> print (V)
```

The output is shown as follows:

```
[[ 1  0  1 -1]
 [-1  0 -1  0]
 [ 1  1  1 -1]
 [ 1  1 -1  0]
 [ 0  0  1 -1]]
```

Let's take a look at the following `pdist` command:

```
>>> pdist(V,metric='cityblock')
```

The output is shown as follows:

```
array([ 5.,  1.,  4.,  1.,  6.,  3.,  4.,  3.,  2.,  5.])
```

This means, if $v1 = [1,0,1,-1]$, $v2 = [-1,0,-1,0]$, $v3 = [1,1,1,-1]$, $v4 = [1,1,-1,0]$, and $v5 = [0,0,1,-1]$, then the Manhattan distance of $v1$ from $v2$ is 5. The distance from $v1$ to $v3$ is 1; from $v1$ to $v4$, 4; and from $v1$ to $v5$, 1. From $v2$ to $v3$ the distance is 6; from $v2$ to $v4$, 3; and from $v2$ to $v5$, 4. From $v3$ to $v4$ the distance is 3; and from $v3$ to $v5$, 2. And finally, the distance from $v4$ to $v5$ is 5, which is the last entry of the output.

- **Third warning**: When computing the distance between each pair of two collections of inputs, we should use the `cdist` routine, which has a similar syntax. For instance, for the two collections of three randomly selected four-dimensional Boolean vectors, the corresponding Jaccard-Needham dissimilarities are computed, as follows:

```
>>> from scipy.spatial.distance import cdist
>>> V=scipy.stats.randint.rvs(0.4, 2, size=(3,4)).astype(bool)
>>> W=scipy.stats.randint.rvs(0.4, 3, size=(2,4)).astype(bool)
>>> cdist(V,W,'jaccard')
array([[ 0.75      ,  1.        ],
       [ 0.75      ,  1.        ],
       [ 0.33333333,  0.5       ]])
```

That is, if the three vectors in `V` are labeled `v1`, `v2`, and `v3` and if the two vectors in `W` are labeled as `w1` and `w2`, then the dissimilarity between `v1` and `w1` is 0.75; between `v1` and `w2`, 1; and so on.

- **Fourth warning**: When we have a large amount of data points and we need to address the problem of nearest neighbors (for example, to locate the closest element of the data to a new instance point), we seldom do it by brute force. The optimal algorithm to perform this search is based on the idea of k-dimensional trees. SciPy has two classes to handle these objects – `KDTree` and `cKDTree`. The latter is a subset of the former, a little faster since it is wrapped from C code, but with very limited use. It only has the `query` method to find the nearest neighbors of the input. The syntax is simple, as follows:

```
KDTree(data, leafsize=10)
```

This creates a structure containing a binary tree, very apt for the design of fast search algorithms. The `leafsize` option indicates at what level the search based on the structure of the binary tree must be abandoned in favor of brute force.

The other methods associated with the `KDTree` class are — `count_neighbors`, to compute the number of nearby pairs that can be formed with another `KDTree`; `query_ball_point`, to find all points at a given distance from the input; `query_ball_tree` and `query_pairs`, to find all pairs of points within certain distance; and `sparse_distance_matrix`, that computes a sparse matrix with the distances between two `KDTree` classes.

Let us see it in action, with a small dataset of 10 randomly generated four-dimensional points with integer entries:

```
>>> from scipy.spatial import KDTree
>>> data=scipy.stats.randint.rvs(0.4,10,size=(10,4))
>>> print (data)
```

The output is shown as follows:

```
[[8 6 1 1]
 [2 9 1 5]
 [4 8 8 9]
 [2 6 6 4]
 [4 1 2 1]
 [3 8 7 2]
 [1 1 3 6]
 [5 2 1 5]
 [2 5 7 3]
 [6 0 6 9]]
>>> tree=KDTree(data)
>>> tree.query([0,0,0,0])
```

The output is shown as follows:

```
(4.6904157598234297, 4)
```

This means, among all the points in the dataset, the closest one in the Euclidean distance to the origin is the fifth one (index 4), and the distance is precisely about 4.6 units.

We can have an input of more than one point; the output will still be a tuple, where the first entry is an array that indicates the smallest distance to each of the input points. The second entry is another array that indicates the indices of the nearest neighbors.

Clustering

Another technique used in data mining is clustering. SciPy has two modules to deal with any problem in this field, each of them addressing a different clustering tool — `scipy.cluster.vq` for k-means and `scipy.cluster.hierarchy` for hierarchical clustering.

Vector quantization and k-means

We have two routines to divide data into clusters using the k-means technique—
kmeans and kmeans2. They correspond to two different implementations.
The former has a very simple syntax:

```
kmeans(obs, k_or_guess, iter=20, thresh=1e-05)
```

The obs parameter is an ndarray with the data we wish to cluster. If the dimensions
of the array are *m* x *n*, the algorithm interprets this data as *m* points in the
n-dimensional Euclidean space. If we know the number of clusters in which this
data should be divided, we enter so with the k_or_guess option. The output is a
tuple with two elements. The first is an ndarray of dimension *k* x *n*, representing
a collection of points—as many as clusters were indicated. Each of these locations
indicates the centroid of the found clusters. The second entry of the tuple is a
floating-point value indicating the distortion between the passed points, and the
centroids generated previously.

If we wish to impose an initial guess for the centroids of the clusters, we may do so
with the k_or_guess parameter again, by sending a *k* x *n* ndarray.

The data we pass to kmeans need to be normalized with the whiten routine.

The second option is much more flexible, as its syntax indicates:

```
kmeans2(data, k, iter=10, thresh=1e-05,
minit='random', missing='warn')
```

The data and k parameters are the same as obs and k_or_guess, respectively. The
difference in this routine is the possibility of choosing among different initialization
algorithms, hence providing us with the possibility to speed up the process and use
fewer resources if we know some properties of our data. We do so by passing to
the minit parameter, one of the strings such as 'random' (initialization centroids
are constructed randomly using a Gaussian), 'points' (initialization is done
by choosing points belonging to our data), or 'uniform' (if we prefer uniform
distribution to Gaussian).

In case we would like to provide the initialization centroids ourselves with the k
parameter, we must indicate our choice to the algorithm by passing 'matrix' to the
minit option as well.

In any case, if we wish to classify the original data by assigning to each point the
cluster to which it belongs; we do so with the vq routine (for vector quantization).
The syntax is pretty simple as well:

```
vq(obs, centroids)
```

The output is a tuple with two entries. The first entry is a one-dimensional `ndarray` of size *n* holding for each point in `obs`, the cluster to which it belongs. The second entry is another one-dimensional `ndarray` of the same size, but containing floating-point values indicating the distance from each point to the centroid of its cluster.

Let us illustrate with a classical example, the mouse dataset. We will create a big dataset with randomly generated points in three disks, as follows:

```
>>> import numpy
>>> from scipy.stats import norm
>>> from numpy import array,vstack
>>> data=norm.rvs(0,0.3,size=(10000,2))
>>> inside_ball=numpy.hypot(data[:,0],data[:,1])<1.0
>>> data=data[inside_ball]
>>> data = vstack((data, data+array([1,1]),data+array([-1,1])))
```

Once created, we will request the data to be separated into three clusters:

```
>>> from scipy.cluster.vq import *
>>> centroids, distortion = kmeans(data,3)
>>> cluster_assignment, distances = vq(data,centroids)
```

Let us present the results:

```
>>> from matplotlib.pyplot import plot
>>> import matplotlib.pyplot as plt
>>> plt.plot(data[cluster_assignment==0,0], \
       data[cluster_assignment==0,1], 'ro')
>>> plt.plot(data[cluster_assignment==1,0], \
       data[cluster_assignment==1,1], 'b+')
>>> plt.plot(data[cluster_assignment==2,0], \
       data[cluster_assignment==2,1], 'k.')
>>> plt.show()
```

This gives the following plot showing the mouse dataset with three clusters from left to right—red (squares), blue (pluses), and black (dots):

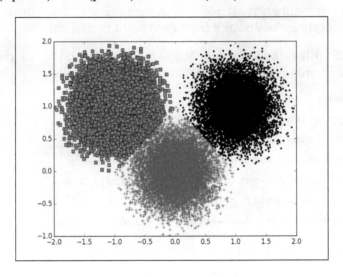

Hierarchical clustering

There are several different algorithms to perform hierarchical clustering. SciPy has routines for the following methods:

- **Single/min/nearest method**: `single`
- **Complete/max/farthest method**: `complete`
- **Average/UPGMA method**: `average`
- **Weighted/WPGMA method**: `weighted`
- **Centroid/UPGMC method**: `centroid`
- **Median/WPGMC method**: `median`
- **Ward's linkage method**: `ward`

In any of the previous cases, the syntax is the same; the only input is the dataset, which can be either an m x n `ndarray` representing m points in the n-dimensional Euclidean space, or a condensed distance matrix obtained from the previous data using the `pdist` routine from `scipy.spatial`. The output is always an `ndarray` representing the corresponding linkage matrix of the clustering obtained.

Alternatively, we may call the clustering with the generic routine `linkage`. This routine accepts a dataset/distance matrix, and a string indicating the method to use. The strings coincide with the names introduced. The advantage of `linkage` over the previous routines is that we are also allowed to indicate a different metric than the usual Euclidean distance. The complete syntax for `linkage` is then as follows:

```
linkage(data, method='single', metric='euclidean')
```

Different statistics on the resulting linkage matrices may be performed with the routines such as Cophenetic distances between observations (`cophenet`); inconsistency statistics (`inconsistent`); maximum inconsistency coefficient for each non-singleton cluster with its descendants (`maxdists`); and maximum statistic for each non-singleton cluster with its descendants (`maxRstat`).

It is customary to use binary trees to represent linkage matrices, and the `scipy.cluster.hierachy` submodule has a large number of different routines to manipulate and extract information from these trees. The most useful of these routines is the visualization of these trees, often called dendrograms. The corresponding routine in SciPy is dendrogram, and has the following imposing syntax:

```
dendrogram(Z, p=30, truncate_mode=None, color_threshold=None,
get_leaves=True, orientation='top', labels=None,
count_sort=False, distance_sort=False,
show_leaf_counts=True, no_plot=False, no_labels=False,
color_list=None, leaf_font_size=None,
leaf_rotation=None, leaf_label_func=None,
no_leaves=False, show_contracted=False,
link_color_func=None)
```

The first obvious parameter, `Z`, is a linkage matrix. This is the only non-optional variable. The other options control the style of the output (colors, labels, rotation, and so on), and since they are technically nonmathematical in nature, we will not explore them in detail in this monograph, other than through the simple application to animal clustering shown next.

Clustering mammals by their dentition

Mammals' teeth are divided into four groups such as incisors, canines, premolars, and molars. The dentition of several mammals has been collected, and is available for download at http://www.uni-koeln.de/themen/statistik/data/cluster/dentitio.dat.

This file presents the name of the mammal, together with the number of top incisors, bottom incisors, top canines, bottom canines, top premolars, bottom premolars, top molars, and bottom molars.

We wish to use hierarchical clustering on that dataset to assess which species are closer to each other by these features.

We will start by preparing the dataset and store the relevant data in ndarrays. The original data is given as a text file, where each line represents a different mammal. The first four lines are as follows:

```
OPOSSUM                     54113344
HAIRY TAIL MOLE             33114433
COMMON MOLE                 32103333
STAR NOSE MOLE             33114433
```

The first 27 characters of each line hold the name of the animal. The characters in positions 28 to 35 are the number of respective kinds of dentures. We need to prepare this data into something that SciPy can handle. We will collect the names apart, since we will be using them as labels in the dendrogram. The rest of the data will be forced into an array of integers:

```python
>>> import numpy
>>> file=open("dentitio.dat","r") # open the file
>>> lines=file.readlines() # read each line in memory
>>> file.close() # close the file
>>> mammals=[] # this stores the names
>>> dataset=numpy.zeros((len(lines),8)) # this stores the data
>>> for index,line in enumerate(lines):
        mammals.append( line[0:27].rstrip(" ").capitalize() )
        for tooth in range(8):
            dataset[index,tooth]=int(line[27+tooth])
```

We will proceed to compute the linkage matrix and its posterior dendrogram, making sure to use the Python list, mammals, as labels:

```python
>>> import matplotlib.pyplot as plt
>>> from scipy.cluster.hierarchy import linkage, dendrogram
>>> Z=linkage(dataset)
>>> dendrogram(Z, labels=mammals, orientation="right")
>>> matplotlib.pyplot.show()
>>> plt.show()
```

This gives us the following dendrogram showing clustering of mammals according to their dentition:

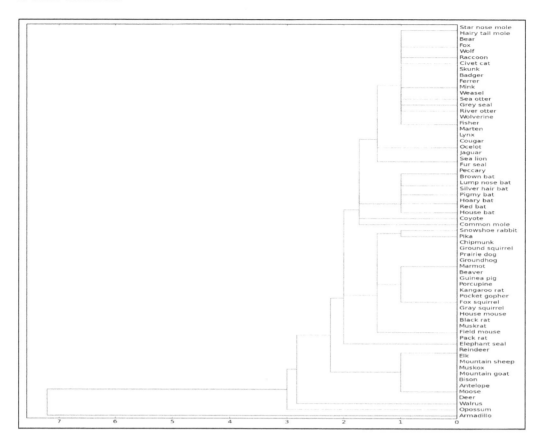

Note how all the bats are clustered together. The mice are also clustered together, but far from the bats. Sheep, goats, antelopes, deer, and moose have similar dentures too, and they appear clustered at the bottom of the tree, next to the opossum and the armadillo. Note how all felines are also clustered together, on the top of the tree.

Experts in data analysis can obtain more information from dendrograms; they are able to interpret the lengths of the branches or the different colors used in the composition, and give us more insightful explanations about the way the clusters differ from each other.

Summary

This chapter dealt with tools appropriate for data mining and explored modules such as `stats` (for statistics), `spatial` (for data structures), and `cluster` (for clustering and vector quantization). In the next chapter, additional functionalities included in the SciPy module, `scipy.spatial`, will be studied, complementing the ones already explored in previous chapters. As usual, each function introduced will be illustrated via non-trivial examples which can be enriched modifying the IPython Notebook corresponding to this chapter.

7
SciPy for Computational Geometry

In this chapter, we will be covering the fundamentals of SciPy to develop programs in this very specialized topic: **Computational Geometry**. Two examples will be used to illustrate the use of SciPy functions in this area. To be able to profit from the first example, you might want to have handy a copy of *Computational Geometry: Algorithms and Applications Third Edition, de Berg M., Cheong O., van Kreveld M.,* and *Overmars M., Springer Publishing*. The second example, on which the **Finite Element Method** is used to solve a two-dimensional problem involving the numerical solution of the Laplace Equation, could be followed without trouble with knowledge on the topic described in *Introduction to the Finite Element Method, Ottosen N. S.* and *Petersson H., Prentice Hall*.

Let's start by covering the routines in the `scipy.spatial` module that deal with the construction of triangulations of points in spaces of any dimension, and the corresponding convex hulls.

The procedure is simple; given a set of m points in the n-dimensional space (which we represent as an m x n NumPy array), we create the `scipy.spatial` class `Delaunay`, containing a triangulation formed by those points:

```
>>> import scipy.stats
>>> import scipy.spatial
>>> data = scipy.stats.randint.rvs(0.4,10,size=(10,2))
>>> triangulation = scipy.spatial.Delaunay(data)
```

Any `Delaunay` class has the basic search attributes such as `points` (to obtain the set of points in the triangulation), `vertices` (that offer the indices of vertices forming simplices in the triangulation), `neighbors` (for the indices of neighbor simplices of each simplex—with the convention that "-1" indicates no neighbor for simplices at the boundary).

More advanced attributes, for example, `convex_hull`, indicate the indices of the vertices that form the convex hull of the given points. If we desire to search for the simplices that share a given vertex, we may do so with the `vertex_to_simplex` method. If, instead, we desire to locate the simplices that contain any given point in the space, we do so with the `find_simplex` method.

At this stage we would like to point out the intimate relationship between triangulations and Voronoi diagrams, and offer a simple coding exercise. Let us start by first choosing a random set of points, and obtaining the corresponding triangulation:

```
>>> from numpy.random import RandomState
>>> rv = RandomState(123456789)
>>> locations = rv.randint(0, 511, size=(2,8))
>>> triangulation=scipy.spatial.Delaunay(locations.T)
```

We may use the `matplotlib.pyplot` routine `triplot` to obtain a graphical representation of this triangulation. We first need to obtain the set of computed simplices. `Delaunay` offers us this set, but by means of the indices of the vertices instead of their coordinates. We, thus, need to map these indices to actual points before feeding the set of simplices to the `triplot` routine:

```
>>> import matplotlib.pyplot as plt
>>> assign_vertex = lambda index: triangulation.points[index]
>>> triangle_set = map(assign_vertex, triangulation.vertices)
```

We will now obtain the edge map of the Voronoi diagram in a similar fashion as we did before (this time using the `scipy.spatial.Voronoi` module), and plot it together with the triangulation. This is done by the following lines of code:

```
>>> voronoiSet=scipy.spatial.Voronoi(locations.T)
>>> scipy.spatial.voronoi_plot_2d(voronoiSet)
>>> fig = plt.figure()
>>> thefig = plt.subplot(1,1,1)
>>> scipy.spatial.voronoi_plot_2d(voronoiSet, ax=thefig)
>>> plt.triplot(locations[1], locations[0], triangles=triangle_set, color='r')
```

Let's take a look at the following `xlim()` command:

```
>>> plt.xlim((0,550))
```

The output is shown as follows:

```
 (0, 550)
```

Now, let's take a look at following `ylim()` command:

```
>>> plt.ylim((0,550))
```

The output is shown as follows:

```
 (0, 550)
```

We now plot the edge map of the Voronoi diagram together with triangulation in the following `plt.show()` command:

```
>>> plt.show()
```

The output is shown as follows:

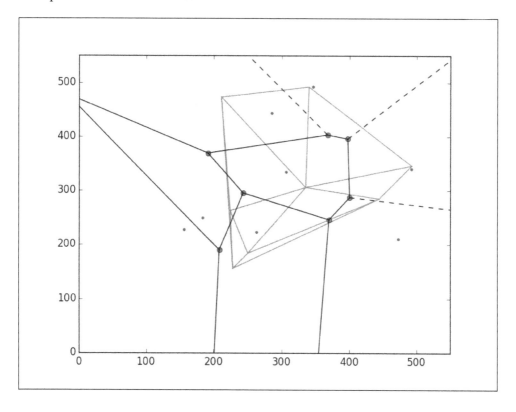

Note how the triangulation and the corresponding Voronoi diagrams are dual of each other; each edge in the triangulation (red) is perpendicular with an edge in the Voronoi diagram (white). How should we use this observation to code an actual Voronoi diagram for a cloud of points? The actual Voronoi diagram is the set of vertices and edges that composes it.

 Interesting ways to find the Voronoi diagram can be found at `http://stackoverflow.com/questions/10650645/python-calculate-voronoi-tesselation-from-scipys-delaunay-triangulation-in-3d`.

Let us finish this chapter with two applications of scientific computing that use these techniques extensively, in combination with routines from other SciPy modules.

The structural model of oxides

In this example, we will cover the extraction of the structural model of a molecule of a bronze-type **Niobium oxide**, from **HAADF-STEM** micrographs (further background on this topic can be found in *Chapter 5, High-Quality Image Formation by Nonlocal Means Applied to High-Angle Annular Dark-Field Scanning Transmission Electron Microscopy (HAADF--STEM) of the book Modeling Nanoscale Imaging in Electron Microscopy, Vogt T., Dahmen W.,* and *Binev P., Springer Publishing.*

The following diagram shows the HAADF-STEM micrograph of a bronze-type Niobium oxide (taken from `http://www.microscopy.ethz.ch/BFDF-STEM.htm`):

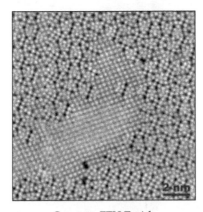

Courtesy: ETH Zurich

For pedagogical purposes, we took the following approach to solving this problem:

- Segmentation of the atoms by thresholding and morphological operations.

- Connected component labeling to extract each single atom for posterior examination.

- Computation of the centers of mass of each label identified as an atom. This presents us with a lattice of points in the plane that shows a first insight in the structural model of the oxide.

- Computation of the Voronoi diagram of the previous lattice of points. The combination of information with the output of the previous step will lead us to a decent (approximation of the actual) structural model of our sample.

Let us proceed in this direction.

Once retrieved and saved in the current working directory, our HAADF-STEM images will be read in python and stored by default (depending on your computer architecture) as big matrices with `float32` or `float64` precision. For this project, it is enough to retrieve some tools from the `scipy.ndimage` module, and some procedures from the `matplotlib` library. The preamble then looks like the following code:

```
>>> import numpy
>>> import scipy
>>> from scipy.ndimage import *
>>> from scipy.misc import imfilter
>>> import matplotlib.pyplot as plt
>>> import matplotlib.cm as cm
```

The image is loaded with the `imread(filename)` command. This stores the image as a `numpy.array` with `dtype = float32`. Notice that the image is rescaled so that the maxima and minima are `1.0` and `0.0`, respectively. Other interesting information about the image can be retrieved as follows:

```
>>> img=imread('./NbW-STEM.png')
>>> minVal = numpy.min(img)
>>> maxVal = numpy.max(img)
>>> img = (1.0/(maxVal-minVal))*(img - minVal)
>>> plt.imshow(img, cmap = cm.Greys_r)
>>> plt.show()
>>> print "Image dtype: %s"%(img.dtype)
>>> print "Image size: %6d"%(img.size)
>>> print "Image shape: %3dx%3d"%(img.shape[0],img.shape[1])
>>> print "Max value %1.2f at pixel %6d"%(img.max(),img.argmax())
```

```
>>> print "Min value %1.2f at pixel %6d"%(img.min(),img.argmin())
>>> print "Variance: %1.5f\nStandard deviation: \
    %1.5f"%(img.var(),img.std())
```

This provides the following output:

```
Image dtype: float64
Image size:  87025
Image shape: 295x295
Max value 1.00 at pixel  75440
Min value 0.00 at pixel   5703
Variance: 0.02580
Standard deviation: 0.16062
```

We perform thresholding by imposing an inequality in the array holding the data. The output is a Boolean array where `True` (white) indicates that the inequality has been fulfilled, and `False` (black) otherwise. We may perform at this point several thresholding operations and visualize them to obtain the best threshold for segmentation purposes. The following images show several examples (different thresholdings applied to the oxide image):

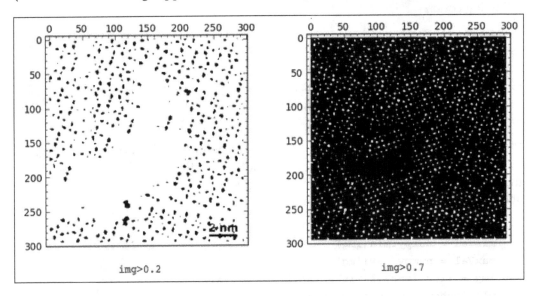

The following lines of code generate that oxide image:

```
>>> plt.subplot(1, 2, 1)
>>> plt.imshow(img > 0.2, cmap = cm.Greys_r)
>>> plt.xlabel('img > 0.2')
```

```
>>> plt.subplot(1, 2, 2)
>>> plt.imshow(img > 0.7, cmap = cm.Greys_r)
>>> plt.xlabel('img > 0.7')
>>> plt.show()
```

By visual inspection of several different thresholds, we choose 0.62 as one that gives us a good map showing what we need for segmentation. We need to get rid of *outliers*, though: small particles that might fulfill the given threshold but are small enough not to be considered as actual atoms. Therefore, in the next step we perform a morphological operation of opening to get rid of those small particles. We decided that anything smaller than a square of size 2 x 2 is to be eliminated from the output of thresholding:

```
>>> BWatoms = (img> 0.62)
>>> BWatoms = binary_opening(BWatoms,structure=numpy.ones((2,2)))
```

We are ready for segmentation, which will be performed with the `label` routine from the `scipy.ndimage` module. It collects one slice per segmented atom and offers the number of slices computed. We need to indicate the connectivity type. For instance, in the following toy example, do we want to consider that situation as two atoms or one atom?

1	1	1	1	1	0
1	1	1	1	0	0
1	1	1	0	0	0
0	0	0	1	1	1
0	0	0	1	1	1
0	0	0	1	1	1

It depends; we would rather have it now as two different connected components, but for some other applications we might consider that they are one. The way we indicate the connectivity to the `label` routine is by means of a structuring element that defines feature connections. For example, if our criterion for connectivity between two pixels is that their edges are adjacent, then the structuring element looks like the image shown on the left-hand side from the images shown next. If our criterion for connectivity between two pixels is that they are also allowed to share a corner, then the structuring element looks like the image on the right-hand side.

For each pixel we impose the chosen structuring element and count the intersections; if there are no intersections, then the two pixels are not connected. Otherwise, they belong to the same connected component.

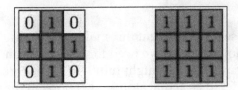

We need to make sure that atoms that are too close diagonally are counted as two, rather than one, so we chose the structuring element on the left. The script then reads as follows:

```
>>> structuring_element = [[0,1,0],[1,1,1],[0,1,0]]
>>> segmentation,segments = label(BWatoms,structuring_element)
```

The segmentation object contains a list of slices, each with a Boolean matrix containing each of the found atoms of the oxide. We may obtain a great deal of useful information for each slice. For example, the coordinates of the center of mars (centers_of_mass) of each atom can be retrieved with the following commands:

```
>>> coords = center_of_mass(img, segmentation, range(1,segments+1))
>>> xcoords = numpy.array([x[1] for x in coords])
>>> ycoords = numpy.array([x[0] for x in coords])
```

Note that because of the way matrices are stored in memory, there is a transposition of the x and y coordinates of the locations of the pixels. We need to take this into account.

Notice the overlap of the computed lattice of points over the original image (the left-hand side image from the two images shown next). We may obtain it with the following commands:

```
>>> plt.imshow(img, cmap = cm.Greys_r)
>>> plt.axis('off')
>>> plt.plot(xcoords,ycoords,'b.')
>>> plt.show()
```

We have successfully found the centers of mass for most atoms, although there are still about a dozen regions where we are not too satisfied with the result. It is time to fine-tune by the simple method of changing the values of some variables; play with the threshold, with the structuring element, with different morphological operations, and so on. We can even add all the obtained information for a wide range of those variables, and filter out outliers. An example with optimized segmentation is shown, as follows (look at the right-hand side image):

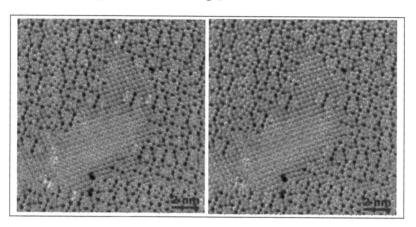

For the purposes of this exposition, we are happy to keep it simple and continue working with the set of coordinates that we have already computed. We will be now offering an approximation to the lattice of the oxide, computed as the edge map of the Voronoi diagram of the lattice:

```
>>> L1,L2 = distance_transform_edt(segmentation==0, return_
distances=False, return_indices=True)
>>> Voronoi = segmentation[L1,L2]
>>> Voronoi_edges= imfilter(Voronoi,'find_edges')
>>> Voronoi_edges=(Voronoi_edges>0)
```

Let us overlay the result of `Voronoi_edges` with the locations of the found atoms:

```
>>> plt.imshow(Voronoi_edges); plt.axis('off'); plt.gray()
>>> plt.plot(xcoords,ycoords,'r.',markersize=2.0)
>>> plt.show()
```

This gives the following output, which represents the structural model we were searching for (recall that we started from an image where we wanted to find the structural model of a molecule):

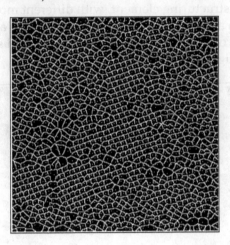

A finite element solver for Laplace's equation

We use finite elements when the size of the data is so large that its results prohibit dealing with finite differences. To illustrate this case, we would like to explore the numerical solution of the Laplace equation, subject to certain boundary conditions.

We will start by defining the computational domain and produce a mesh dividing this domain using triangles as local finite elements. This will be our starting point to solve this problem using finite elements, as we will be placing on the computational domain a piecewise continuous function, whose pieces are linear and supported on each of the triangles.

We start by calling the necessary modules to build the mesh (other modules will be called as they are required):

```
>>> import numpy
>>> from numpy import linspace
>>> import scipy
>>> import matplotlib.pyplot as plt
>>> from scipy.spatial import Delaunay
```

First we define the region:

```
>>> xmin = 0 ; xmax = 1 ; nXpoints = 10
>>> ymin = 0 ; ymax = 1 ; nYpoints = 10
>>> horizontal = linspace(xmin,xmax,nXpoints)
>>> vertical = linspace(ymin,ymax,nYpoints)
>>> y, x = numpy.meshgrid(horizontal, vertical)
>>> vertices = numpy.array([x.flatten(),y.flatten()])
```

We may now create the triangulation:

```
>>> triangulation = Delaunay(vertices.T)
>>> index2point = lambda index: triangulation.points[index]
>>> all_centers = index2point(triangulation.vertices).mean(axis=1)
>>> trngl_set=triangulation.vertices
```

We then have the following triangulation:

```
>>> plt.triplot(vertices[0],vertices[1],triangles=trngl_set)
>>> plt.show()
```

This produces the following graph:

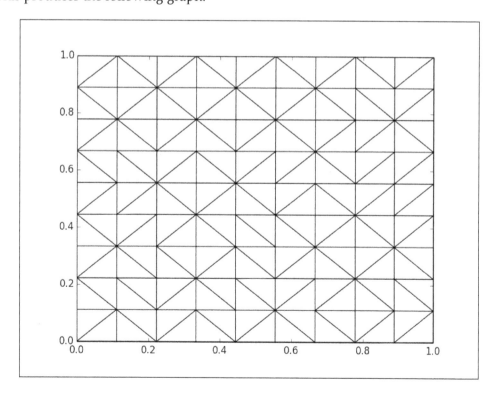

In this case, the problem we have chosen is a standard one in mathematical methods in Physics and Engineering, consisting of solving the two-dimensional Laplace's equation on the unit square region, with zero **Dirichlet** boundary conditions on three sides and, on the fourth side, a constant. Physically, this problem could represent diffusion of temperature on a two-dimensional plate. Mathematically, the problem is formulated in the following form:

$$\begin{cases} \nabla^2\phi(x, y) = 0 \\ \phi(x = 0, y) = 0; \; \phi(x = 1, y) = 1; \; y \neq 0 \text{ and } y \neq 1 \\ \phi(x, y = 0) = \phi(x, y = 1) = 0 \end{cases}$$

The solution of this form can be given in terms of Fourier series as follows:

$$\phi(x, y) = 2\sum_{n=1}^{\infty}\left[\frac{1}{n\pi} - \frac{\cos(n)}{n\pi}\right]\frac{\sinh(n\pi x)}{\sinh(n\pi)}\sin(n\pi y)$$

This is important as you can check the correctness of the obtained numerical solution before attempting to use your numerical scheme to tackle more complex problems in complex computational domains. It should be mentioned, however, that there are alternatives in Python that implement the finite element method to solve partial differential equations. In this regard, the reader could consult the **Fenics** project (`http://fenicsproject.org/book/`) and the **SfePy** project (`http://sfepy.org/doc-devel/index.html`).

We code the solution in the usual fashion. We first compute the stiff matrix A (which for obvious reasons is `sparse`). Then, the construction of the vector, R, holding global boundary conditions is defined (the way we have constructed our mesh makes defining this vector straightforward). With them, the solution to the system comes from the solution X obtained from solving a matrix equation of the form $AX=R$ using a subset of the matrices A and R corresponding to the nodes different from those on the boundaries. This should be no trouble for SciPy. Let us start with the stiff matrix:

```
>>> from numpy import   cross
>>> from scipy.sparse import dok_matrix
>>> points=triangulation.points.shape[0]
```

```
>>> stiff_matrix=dok_matrix((points,points))
>>> for triangle in triangulation.vertices:
        helper_matrix=dok_matrix((points,points))
        pt1,pt2,pt3=index2point(triangle)
        area=abs(0.5*cross(pt2-pt1,pt3-pt1))
        coeffs=0.5*numpy.vstack((pt2-pt3,pt3-pt1,pt1-pt2))/area
        #helper_matrix[triangle,triangle] = \
        array(mat(coeffs)*mat(coeffs).T)
        u=None
        u=numpy.array(numpy.mat(coeffs)*numpy.mat(coeffs).T)
        for i in range(len(triangle)):
            for j in range(len(triangle)):
                helper_matrix[triangle[i],triangle[j]] = u[i,j]
        stiff_matrix=stiff_matrix+helper_matrix
```

Note that this is the cumbersome way to update the matrix stiff_matrix. This is due to the fact that the matrix is sparse, and the current choice of representation does not behave well with indexing.

To compute the global boundary vector we need to collect all edges on the boundary first and then assign to the nodes with $x=1$ that the function is one and to the others that the function is zero. Because of the way we set up the mesh this is easy as the nodes on which the function will take the value of one are always the last entries in the global boundary vector. This is accomplished by the following lines of code:

```
>>> allNodes = numpy.unique(trngl_set)
>>> boundaryNodes = numpy.unique(triangulation.convex_hull)
>>> NonBoundaryNodes = numpy.array([])
>>> for x in allNodes:
        if x not in boundaryNodes:
            NonBoundaryNodes = numpy.append(NonBoundaryNodes,x)
    NonBoundaryNodes = NonBoundaryNodes.astype(int)
    nbnodes = len(boundaryNodes) # number of boundary nodes
    FbVals=numpy.zeros([nbnodes,1]) # Values on the boundary
    FbVals[(nbnodes-nXpoints+1):-1]=numpy.ones([nXpoints-2, 1])
```

We are ready to find the numerical solution to the problem with the values obtained in our previous step:

```
>>> totalNodes = len(allNodes)
>>> stiff_matrixDense = stiff_matrix.todense()
>>> stiffNonb = \
    stiff_matrixDense[numpy.ix_(NonBoundaryNodes,NonBoundaryNodes)]
>>> stiffAtb = \
    stiff_matrixDense[numpy.ix_(NonBoundaryNodes,boundaryNodes)]
>>> U=numpy.zeros([totalNodes, 1])
>>> U[NonBoundaryNodes] = numpy.linalg.solve( - stiffNonb , \
    stiffAtb * FbVals )
>>> U[boundaryNodes] = FbVals
```

This produces the following image depicting the diffusion of temperature inside the square:

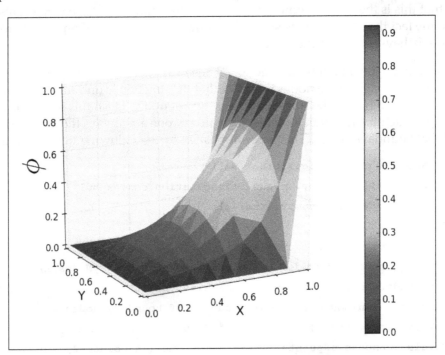

This graph was obtained in the following way:

```
>>> X = vertices[0]
>>> Y = vertices[1]
>>> Z = U.T.flatten()
>>> from mpl_toolkits.mplot3d import axes3d
>>> fig = plt.figure()
>>> ax = fig.add_subplot(111, projection='3d')
>>> surf = ax.plot_trisurf(X, Y, Z, cmap=cm.jet, linewidth=0)
>>> fig.colorbar(surf)
>>> fig.tight_layout()
>>> ax.set_xlabel('X',fontsize=16)
>>> ax.set_ylabel('Y',fontsize=16)
>>> ax.set_zlabel(r"$\phi$",fontsize=36)
>>> plt.show()
```

An important point in numerical analysis is to evaluate the quality of the numerical solution obtained to any problem. In this case, we have chosen a problem whose analytical solution is available (see the preceding code), so one could check (not prove) the validity of the numerical algorithm implemented to solve our problem. In this case the analytical solution can be coded in the following manner:

```
>>> from numpy import pi, sinh, sin, cos, sum
>>> def f(x,y):
        return sum( 2*(1.0/(n*pi) - \
        cos(n*pi)/(n*pi))*(sinh(n*pi*x)/ \
        sinh(n*pi))*sin(n*pi*y)
                for n in range(1,200))
>>> Ze = f(X,Y)
>>> ZdiffZe = Ze - Z
>>> plt.plot(ZdiffZe)
>>> plt.show()
```

This produces the following graph showing the difference between the exact solution (evaluated up to 200 terms) and the numerical solution of the problem (via the corresponding IPython notebook you could perform some further analysis on the numerical solution just to become more confident on the rightness of the obtained result):

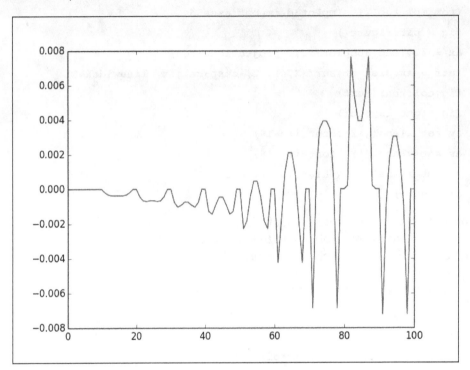

Summary

In each one of the seven chapters of this book, we have covered at length all the different modules included in the SciPy libraries in a structured manner, derived from the logical division of the different branches of mathematics.

We have also witnessed the power of this system to accomplish with minimal coding and optimal resource use, state-of-the-art applications to research problems in different areas of science.

In *Chapter 8, Interaction with Other Languages*, we will introduce one of the main strengths of SciPy: the ability to interact with other languages.

8
Interaction with Other Languages

We often need to incorporate into our workflow some code written in different languages; mostly C/C++ or Fortran, and also from R, MATLAB, or Octave. Python excels at allowing code from all these other sources to run from within; care must be taken to convert different numerical types to something that Python understands, but this is pretty much the only issue we encounter.

If you are working with SciPy, it is because your Python ecosystem has available compilers for C and Fortran programs. Otherwise, SciPy could have not been installed on your system. Also, given its popularity, it is highly probably that your computer environment has MATLAB/Octave available. Accordingly, this has driven the selection of topics listed later in this chapter. We left to the interested reader to find out how interface with R and many other software is available out there for numerical computing. Two alternatives to do that with R are the packages **PypeR** (`http://bioinfo.ihb.ac.cn/softwares/PypeR/`) and **rpy2** (`http://rpy.sourceforge.net/`). Additional alternatives can be found at `http://stackoverflow.com/questions/11716923/python-interface-for-r-programming-language`.

In this chapter, we will cover the following things:

- A brief discussion on how Python can be used to run codes from Fortran, C/C++, and MATLAB/Octave
- We will first see the basic functionality of the utility `f2py` to handle the inclusion of Fortran codes in Python via SciPy
- A basic usage to include C/C++ code within Python code using the tools provided by the the `scipy.weave` module

The routines will be illustrated via simple examples that can be enriched by you modifying the IPython Notebook corresponding to this chapter.

Interaction with Fortran

SciPy provides a simple way of including Fortran code—f2py. This is a utility shipped with the NumPy libraries, which is operative when distutils from SciPy are available. This is always the case when we install SciPy.

The f2py utility is supposed to run outside Python, and it is used to create from any Fortran file a Python module that can be easily called in our sessions. Under any *nix system, we call it from the terminal. Under Windows, we recommend you run it in the native terminal, or even better, through a cygwin session.

Before being compiled with f2py, any Fortran code needs to undergo three basic changes, which are as follows:

- Removal of all allocations

- Transformation of the whole program into a subroutine

- If anything special needs to be passed to f2py, we must add it with the comment string "!f2py" or "cf2py"

Let's illustrate the process with a simple example. The following naive subroutine, which we store in the primefactors.f90 file, performs a factorization in prime numbers for any given integer:

```fortran
SUBROUTINE PRIMEFACTORS(num, factors, f)
  IMPLICIT NONE
  INTEGER, INTENT(IN) :: num   !input number
  INTEGER, INTENT(OUT), DIMENSION((num/2))::factors
  INTEGER, INTENT(INOUT) :: f
  INTEGER :: i, n
  i = 2
  f = 1
  n = num
  DO
    IF (MOD(n,i) == 0) THEN
      factors(f) = i
      f = f+1
      n = n/i
    ELSE
      i = i+1
    END IF
    IF (n == 1) THEN
```

```
      f = f-1
    EXIT
  END IF
END DO
```

Since no allocation was made in the code, and we receive a subroutine directly, we may skip to the third step, but for the moment we will not tamper with f2py commands, and are content with trying to create a python module from it. The fastest way to wrap this primefactors subroutine is by issuing the following command (at the shell or terminal prompt indicated by %):

```
% f2py -c primefactors.f90 -m primefactors
```

If everything is correct, an extension module with the name primefactors. so is created. We can then access the primefactors routine in Python from the primefactors module:

```
>>> import primefactors
>>> primefactors.primefactors(6,1)
```

The output is shown as follows:

```
array([2, 3, 0], dtype=int32)
```

Interaction with C/C++

Technically, f2py can also wrap a C code for us, but there are more efficient ways to perform this task. For instance, if we need to interface a very large library of C functions, the preferred method for doing this is **Simplified Wrapper and Interface Generator (SWIG)** (http://www.swig.org/). To wrap C++ code, depending on the features required and the method of interacting with Python, we have several methods such as SWIG or f2py again, but also **PyCXX**, **Boost.Python**, **Cython**, or the SciPy module: weave. When C compilers are not available (and thus linking extensive libraries is not possible in the usual way), we use ctypes. Whenever we will use NumPy/SciPy code, and want fast solutions to our wrapping/binding, the two most common ways to interact with C/C++ are usually through the Python/C API and weave packages.

All the methods briefly enumerated here would require an entire monograph to describe, at length, the methodology of binding the nuisances of the wrapping, depending on systems and requirements, and the caveats of their implementations. The method we would like to cover in more detail in this chapter is the weave package, more concretely by means of the inline routine. This command receives a string (raw or otherwise) containing a sequence of commands, and runs it in Python by calling your C/C++ compiler. The syntax is as follows:

```
inline(code, arg_names, local_dict=None, global_dict=None,
            force = 0,
            compiler='',
            verbose = 0,
    support_code = None,
            customize=None,
    type_factories = None,
    auto_downcast=1,
            **kw)
```

Let's go over the different parameters:

- The `code` parameter is the string that holds the code to be run. Note that this code must not specify any kind of `return` statement. Instead, it should assign some result that can be returned to Python.

- The `arg_names` parameter is a list of strings containing the Python variable names that are to be sent to the C/C++ code.

- The `local_dict` parameter is optional, and must be a Python dictionary containing the values used as local scope for the C/C++ code.

- The `global_dict` parameter is also optional, and must be another Python dictionary containing the values that should be used as the global scope for the C/C++ code.

- The `force` parameter is used only for debugging purposes. It is also optional, and can take only two values — 0 (by default) or 1. If its value is set to 1, the C/C++ code is compiled every time `inline` is called.

- We may specify the compiler that takes over the C/C++ code with the `compiler` option. It must be a string containing the name of the C/C++ compiler.

Let's take an example of the `inline` routine in which we use the following method to employ `cout` for text displaying purposes:

```
>>> import scipy.weave
>>> name = 'Francisco'
>>> pin = 1234
>>> code = 'std::cout << name << "---PIN: " '
>>> code+= '<<std::hex << pin <<std::endl;'
>>> arg_names = ['name','pin']
>>> scipy.weave.inline(code, arg_names)
```

The output is shown as follows:

```
Francisco---PIN: 4d2
```

That was a very simple example, in which no external header declarations were needed. If we wish to do so, those go into the `support_code` option. For instance, if we wish to include math functions from R in our C/C++ code and pass it with `inline`, we need to perform the following steps:

1. Configure the C functions as a shared library. In the folder, holding the R release in a terminal session, issue the following command:

   ```
   % ./configure --enable-R-static-lib --enable-static --with
   -readline=no
   ```

2. Change to the `standalone` folder at `src/nmath` and finish the installation of the libraries. At the end, we should have a file named `libRmath.so`, which needs to be pointed to from the `libpath` string back into our Python session:

   ```
   % cd src/nmath/standalone
   % make
   ```

3. Back in our Python session, we prepare the `inline` call with the proper options. For instance, if we wish to call the R routine `pbinom`, we proceed as follows:

   ```
   >>> import scipy.weave
   >>> support_code= 'extern "C" double pbinom(double x,\
                       double n, double p, int lower_tail, int log_p);'
   >>> libpath='/opt/Rlib' #IS THE LOCATION OF LIBRARY libRmath.so
   >>> library_dirs=[libpath]
   >>> libraries=['Rmath']
   >>> runtime_library_dirs=[libpath]
   >>> code='return_val=pbinom(100,20000,100./20000.,0,1);'
   >>> res=scipy.weave.inline(code, support_code=support_code, \
           library_dirs=library_dirs, libraries=libraries, \
           runtime_library_dirs=runtime_library_dirs)
   >>> print(res)
   ```

   The output is shown as:

   ```
   -0.747734910363
   ```

Note how the function declaration is passed in `support_code`, not in code. Also, note that this option needs to start with `extern "C"` whenever we are not using C++.

4. If extra headers need to be passed, we do so with the `header` option, rather than `support_code` or code:

```
>>> headers = ['<math.h>']
```

We have a word of advice. Care must be taken while converting the different variable types from their original C/C++ format to something that Python understands. This requires modifying the original C/C++ code in certain cases. But by default, we do not have to worry about the following C/C++ types, as SciPy automatically turns them into the indicated Python formats, as shown in the following table:

Python	int	float	complex	string	list	dict	tuple
C/C++	int	double	std::complex	py::string	py::list	py:dict	py::tuple

File types `FILE*` are sent to Python files. Python callables and instances are both obtained from `py::object`. NumPy ndarrays are constructed from `PyArrayObject*`. For any other Python type to be used, the corresponding C/C++ types must be carefully turned into combinations of the previous.

And that should be all. To go beyond trivial uses of the inline function, we usually create extension modules and catalog the functions within for future use.

Interaction with MATLAB/Octave

Since both numerical computing environments are provide with a fourth-generation programming language, we discourage the straightforward inclusion of code from any of these two. There is no gain in terms of speed, resource usage, or coding power. In the extreme and rare cases, in which a specific routine is not available in SciPy, the preferred way to bring it to our session is by generating C code from the MATLAB/Octave code, and then wrap it with any of the methods suggested in the *Interaction with C/C++* section of this chapter.

There is a different story when we receive data created from within MATLAB or Octave. SciPy has a dedicated module to deal with this situation – `scipy.io`.

Let's show you by example. We start with Octave, where we generate a **Delaunay triangulation** of a random set of 10 points in the plane.

We save the coordinates of these points, as well as the pointers to the triangles in the triangulation, to a MATLAB-style file (version 7) called data:

```
octave:1> x=rand(1,10);
octave:2> y=rand(size(x));
octave:3> T=delaunay(x,y);
octave:4> save -v7 data x y T
```

We are done here. We then go to our Python session, where we recover the file data:

```
>>> from scipy.io import loadmat
>>> datadict = loadmat("data")
```

The `datadict` variable holds a Python dictionary with the names of the variables as keys and the loaded matrices as their corresponding values:

```
>>> datadict.keys()
```

The output is shown as follows:

```
['__header__', '__globals__', 'T', 'y', 'x', '__version__']
```

Let's issue the `datadict` command:

```
>>> datadict['x']
```

The output is shown as follows:

```
array([[0.81222999,0.51836246,0.60425982,0.23660352,0.01305779,
        0.0875166,0.77873049,0.70505801,0.51406693,0.65760987]])
```

Let's take a look at following `datadict` command:

```
>>> datadict['__header__']
```

The output is shown as follows:

```
'MATLAB 5.0 MAT-file, written by Octave 3.2.4, 2012-11-27
 15:45:20 UTC'
```

It is possible to save data from our sessions to a format that MATLAB and Octave will understand. We do so with the `savemat` command, from the same module. The syntax is as follows:

```
savemat(file_name, mdict, appendmat=True, format='5',
long_field_names=False, do_compression=False,
oned_as=None)
```

The file_name parameter contains the name of the MATLAB-type file where the data will be written. The Python dictionary mdict contains the names (as keys) of the variables, and their corresponding array values.

If we wish to append .mat at the end of the file, we may do so in the file_name variable, or by setting appendmat to True. In case we need to provide long names for the files (which not all versions of MATLAB accept), we need to indicate so by setting the long_field_names option to True.

We may indicate the version of MATLAB with the format option. We set it to the string '5' for versions 5 and later, or to the string '4' for version 4.

It is possible to compress the matrices we send, and we indicate so by setting the do_compression option to True.

The last option is very interesting. It allows us to indicate to MATLAB/Octave whether our arrays are to be read column by column, or row by row. Setting the oned_as parameter to the string 'column' will send our data into a collection of column vectors. If we set it to the string 'row', it will send the data as collections of row vectors. If set to None, the format in which the data was written is respected.

Summary

This chapter introduced one of the main strengths of SciPy—the ability to interact with other languages such as C/C++, Fortran, R, and MATLAB/Octave. To go in depth into interfacing Python with other languages, you might want to read more specialized literature like *Learning Cython Programming, Philip Herron, Packt Publishing* or the in-depth coverage of F2PY at http://docs.scipy.org/doc/numpy/f2py/ and http://www.f2py.com/home/references. Additional help can be found at https://wiki.python.org/moin/IntegratingPythonWithOtherLanguages.

If you have reached this chapter and have been reading from the first one, you should be aware that many topics were left out in this introductory chapter on SciPy. This book has given you enough background to further strengthen your skills and ability to work with SciPy. To proceed studying, refer to the SciPy Reference Guide (http://docs.scipy.org/doc/scipy/reference/) and other documentation guides available at (http://docs.scipy.org/doc/).

In addition, we recommend you regularly read and also subscribe to the SciPy mailing list (`http://mail.scipy.org/mailman/listinfo/scipy-user`) where you can interact with users of SciPy all over the world, not only by asking/answering questions about SciPy, but also to find out current trends on SciPy and even jobs related to it.

You can peruse the historical archive of the collection of postings to the list, `http://mail.scipy.org/pipermail/scipy-user/`. Also, you should know that there is a SciPy conference held every year (`http://conference.scipy.org/`) which, to quote them, allows participants from academic, commercial, and governmental organizations to showcase their latest Scientific Python projects, learn from skilled users and developers, and collaborate on code development.

Index

E

Eigenvalue problems 60, 61
elliptic integrals 96
empty command 33
exponential integral 95

F

filters
about 110-113
design 114, 115
image interpolation 116-119
LTI system theory 113, 114
morphology 119-121
finite element solver
for Laplace's equation 152-157
Finite impulse response (FIR) 114
Fortran
interacting with 160, 161
functions
Airy function 76, 77
Bairy function 76
Bessel function 77
elliptic functions 78
evaluating 67
gamma function 74, 75
Riemann zeta function 75, 76
special functions 78
Struve function 77

G

gamma function 74, 75
gamma integral 96
GNU Octave system 8

H

HAADF-STEM
about 146
URL 146
hierarchical clustering 138
HippoDraw
URL 9
Horner scheme
URL 67
hyperbolic trigonometric integral 95, 96

I

identity command 33
image compression
via SVD 62
image interpolation 116-119
image processing algorithms 23-25
Infinite impulse response (IIR) 114
integration
about 94
beta integral 96
elliptic integrals 96
exponential integral 95
gamma integral 96
hyperbolic trigonometric integral 95, 96
logarithm integral 95
numerical integration 97, 98
trigonometric integral 95, 96
interpolation 80-85
interval estimation 126-128
IPython Notebook
opening 20, 21
URL 20

K

kernel density estimation 130
Kolmogorov-Smirnov tests 126
KroghInterpolator command 81

L

Laplace's equation
finite element solver, used for 152-157
Levenberg-Marquardt algorithm 87
logarithm integral 95
Lorenz attractors 101-103
LTI system theory 113, 114

M

Mac OS X
SciPy, installing on 11
mammals
clustering, by dentition 139-141
Maple 8
masking 35
Mathematica 8

Thank you for buying
Learning SciPy for Numerical and Scientific Computing
Second Edition

About Packt Publishing

Packt, pronounced 'packed', published its first book, *Mastering phpMyAdmin for Effective MySQL Management*, in April 2004, and subsequently continued to specialize in publishing highly focused books on specific technologies and solutions.

Our books and publications share the experiences of your fellow IT professionals in adapting and customizing today's systems, applications, and frameworks. Our solution-based books give you the knowledge and power to customize the software and technologies you're using to get the job done. Packt books are more specific and less general than the IT books you have seen in the past. Our unique business model allows us to bring you more focused information, giving you more of what you need to know, and less of what you don't.

Packt is a modern yet unique publishing company that focuses on producing quality, cutting-edge books for communities of developers, administrators, and newbies alike. For more information, please visit our website at www.packtpub.com.

About Packt Open Source

In 2010, Packt launched two new brands, Packt Open Source and Packt Enterprise, in order to continue its focus on specialization. This book is part of the Packt Open Source brand, home to books published on software built around open source licenses, and offering information to anybody from advanced developers to budding web designers. The Open Source brand also runs Packt's Open Source Royalty Scheme, by which Packt gives a royalty to each open source project about whose software a book is sold.

Writing for Packt

We welcome all inquiries from people who are interested in authoring. Book proposals should be sent to author@packtpub.com. If your book idea is still at an early stage and you would like to discuss it first before writing a formal book proposal, then please contact us; one of our commissioning editors will get in touch with you.

We're not just looking for published authors; if you have strong technical skills but no writing experience, our experienced editors can help you develop a writing career, or simply get some additional reward for your expertise.

Learning SciPy for Numerical and Scientific Computing

ISBN: 978-1-78216-162-2   Paperback: 150 pages

A practical tutorial that guarantees fast, accurate, and easy-to-code solutions to your numerical and scientific computing problems with the power of Scipy and Python

1. Perform complex operations with large matrices, including eigenvalue problems, matrix decompositions, or solution to large systems of equations.

2. Step-by-step examples to easily implement statistical analysis and data mining that rivals in performance any of the costly specialized software suites.

Learning NumPy Array

ISBN: 978-1-78398-390-2   Paperback: 164 pages

Supercharge your scientific Python computations by understanding how to use the NumPy library effectively

1. Improve the performance of calculations with clean and efficient NumPy code.

2. Analyze large data sets using statistical functions and execute complex linear algebra and mathematical computations.

3. Perform complex array operations in a simple manner.

Please check **www.PacktPub.com** for information on our titles

NumPy Beginner's Guide
Second Edition

ISBN: 978-1-78216-608-5   Paperback: 310 pages

An action packed guide using real world examples of the easy to use, high performance, free open source NumPy mathematical library

1. Perform high performance calculations with clean and efficient NumPy code.

2. Analyze large data sets with statistical functions.

3. Execute complex linear algebra and mathematical computations.

NumPy Cookbook

ISBN: 978-1-84951-892-5   Paperback: 226 pages

Over 70 interesting recipes for learning the Python open source mathematical library, NumPy

1. Do high performance calculations with clean and efficient NumPy code.

2. Analyze large sets of data with statistical functions.

3. Execute complex linear algebra and mathematical computations.

Please check **www.PacktPub.com** for information on our titles

9 781783 987702